T0294847

SCIENCE, TECHNOLOGY, AND SOCIETY IN THE THIRD WORLD

an annotated bibliography

by

WESLEY SHRUM
CARL L. BANKSTON III
and
D. STEPHEN VOSS

The Scarecrow Press, Inc.
Metuchen, N.J., & London
1995

British Library Cataloguing-in-Publication data available

Library of Congress Cataloging-in-Publication Data

Shrum, Wesley, 1953-
 Science, technology, and society in the Third World : an annotated bibliography / by Wesley Shrum, Carl L. Bankston, III and D. Stephen Voss
 p. cm.
 Includes bibliographical references and index.
 ISBN 0-8108-2871-5 (acid-free paper)
 1. Science and state--Developing countries--Bibliography.
 2. Science--Social aspects--Developing countries--Bibliography.
 3. Technology and state--Developing countries--Bibliography.
 4. Technology--Social aspects--Developing countries--Bibliography.
 I. Bankston, Carl L. (Carl Leon), 1952- . II. Voss, D. Stephen (Dennis Stephen), 1968- . III. Title.
Z7407.D44S57 1995
[Q127.D44]
338.9'26'091724--dc20 94-6256

Manufactured in the United States of America
Printed on acid-free paper

Table of Contents

Part I -- Introduction

Given the overwhelming and generally recognized importance of science and technology in the world today, scholars, policymakers, and managers have long sought to understand their growth, direction, and effects. The recent success of Taiwan, Singapore, and Korea as producers and exporters of manufactured goods has underscored their importance in less-developed countries, as well as in highly industrialized nations. Still, the prominent role of science and technology in the Third World often suffers neglect, paling beside the highly visible problems of poverty, population growth, and inequality. Our feeling, shared by many of the scholars whose work is annotated here, is that science and technology are critically important to less-developed countries as well. That they have begun to receive scholarly attention is amply illustrated by the contents of this volume.

In late 1989, as part of an effort to understand the role of science and technology in less-developed countries, we began to search the literature for guides to the area. We quickly discovered a lack of comprehensive bibliographies. What sources do exist are partial, and nothing is available which might be considered current. This volume is intended to fill that gap.

Contents

As the table of contents indicates, our coverage is broad and interdisciplinary, although the volume of studies is such that no single work can include all of the publications in the area. We attempted to isolate works in which science and/or technology were a major focus, whether dealing with less-developed countries in general or with specific countries of the Third World. By "the Third World" we mean the less-developed countries of Asia, Africa, Latin America, and the Pacific. (We do not include Eastern European countries.)

We employ a very general and conventional definition of science and technology: research work done in university, governmental, and industrial laboratories; manufacture, sales, and use of goods; social, political, and environmental effects of these products and processes; and in a few cases the educational systems which produce scientists and engineers.

Time Frame

The coverage extends from 1975 to the fall of 1992. Studies published in the late 1970s and early 1980s often report data collected or research completed prior to this starting date. However, coverage is not generally intended to reflect work done prior to the mid 1970s. This bibliography roughly covers a period in which the dependency perspective on development both flourished and declined, leading to the present rather heterogeneous state of affairs.

Coverage Decisions

The bibliography is interdisciplinary, including all of the social sciences (sociology, psychology, economics, anthropology, political science, development studies, etc.). The objective is to provide an entryway into the literature; these annotations are of course not a substitute for reading the works cited. It is a guide to works which should be available readily at any large university library or through the interlibrary loan system. We have excluded dissertations, reports to funding agencies, departmental report series, and works that are not written in English or translated. That is, we focused on annotating works that are relatively easy for the reader to obtain, instead of seeking to provide a summary of works which would be difficult to find in the original versions.

Of course, this decision means that the large volume of science and technology studies which are done in Chinese, for example, as well as those works which are published and distributed in India (even if in English) are not included. These books and essays are unlikely to be available to the reader, and in any case would swell the size of the bibliography unacceptably.

Procedure

We utilized a formal, sequential search procedure rather than beginning with citation searches of key articles. The following steps were used:

(1) We began by searching the January 1, 1990, issue of *Current Contents*, a journal published weekly by the Institute for Scientific Information that includes a semi-annual social and behavioral

sciences edition listing relevant scholarly periodicals. We searched the "journals" and "annuals" listings for three main subjects: development, science and technology, and the Third World.

(2) We initially listed all promising titles (about 350) by discipline and region, only excluding obviously minor or irrelevant sources.

(3) After a preliminary search we deleted the more general journals and those which rarely published anything on the less-developed countries.

(4) We used a table of contents and abstract search of (a) *Social Studies of Science*, (b) *Science, Technology and Human Values*, and (c) *World Development* for two purposes -- first, to get a preliminary estimate of the search time required, and second, to assess the relevance of specific subjects to the overall project. As a result, we decided to exclude more general articles (e.g., in *World Development*) dealing primarily with economic dependence and capital/labor intensiveness that did not focus specifically on science and technology. We decided to include some "intermediate technology" case studies.

(5) We supplemented the above by searching AGRICOLA and ERIC, which are CD-ROM (compact disk) databases with a focus on agriculture and education; the past 15 years of *Sociological Abstracts* (index entries for development, education, science and technology), *Contemporary Sociology* (a book-review journal, section on Development), and the *Social Science Abstracts Index* (same search terms as above).

(6) For very recent articles (since early 1989) we searched *Current Contents* each week until fall 1992 for relevant materials, including the index terms "third world," "developing countries/nations," "less-developed countries," and "technology transfer."

(7) We then examined four regional journals for Latin America and two general social science journals as a completeness check.

(8) We added references from the bibliographies of the main review articles.

Throughout the search, we were amazed at the nooks and crannies of the literature. Although we were committed to including the bulk of the sources identified through this method, we were forced to exclude items during the annotation stage, especially when they were not clearly related to our subject or when an author had several related publications on the same topic. As our database went through some two dozen revisions, we finally settled on 978 articles and books.

Format

The bibliography is divided into two main parts: the first deals more with economics and innovation, the second with policy and political factors. Each part is introduced by a short essay providing an orientation to relevant works. Chapter 1 also may be used as a quick orientation to the literature. Here we included review articles, reference materials, and bibliographies, as well as sources we felt might be suitable for college teaching.

The body of the book groups bibliographic entries by broad subject matter. It is followed by author and subject indices useful for finding entries of interest; nevertheless, simply scanning the chapters often will prove the most productive technique for source finding. The individual entries in each chapter are numbered. These numbers, and not page numbers, are reported in the indices at the back of the book. These numbers also are used as citations in the two essay chapters, so that a researcher can jump straight from an essay to a particular entry it discusses.

The annotations were written to assist researchers in deciding which articles and books are relevant for specific projects. They were not intended as comprehensive summaries of a work's scholarly arguments, although often this was necessary to establish the subject matter. For this reason, some annotations will limit themselves to an outline of the subjects discussed in an article or book. No annotation will serve as an alternative to reading the actual research.

Abbreviations

Although we attempted to avoid jargon and abbreviations whenever possible, the use of both often was necessary for efficiency of presentation. Only a handful of abbreviations were used without explanation. These include:

ASEAN	Association of South East Asian Nations
AT	Appropriate Technology
Benelux	Belgium, the Netherlands and Luxembourg
DOS	A personal computer operating system developed by Microsoft

GATT	General Agreement on Tariffs and Trade
IEEE	Institute of Electrical and Electronics Engineers
MNC	Multinational Corporation
MNE	Multinational Enterprise
NIC	Newly Industrialized Country
OECD	Organization for Economic Cooperation and Development
R&D	Research and Development
S&T	Science and Technology
SIS	Society for Information Science
STD	Science, Technology and Development
STS	Science, Technology and Society
TA	Technology Assessment
TNC	Transnational Corporation
TT	Technology Transfer
UN	United Nations
UNCSTD	UN Conference on Science and Technology for Development
UNCTAD	UN Conference on Trade and Development
UNDP	UN Development Programme
UNESCO	UN Educational, Scientific and Cultural Organization
UNISIST	An information retrieval system
UNRISD	UN Research Institute for Social Development

For a general review of the field, the reader may wish to consult "Science and Technology in Less-Developed Countries," by Wesley Shrum and Yehouda Shenhav, in the *Handbook of Science, Technology, and Society*, edited by James Petersen, Gerald Markle, Sheila Jasanoff, and Trevor Pinch (Newbury Park, CA: Sage Publications, 1994).

Wesley Shrum

Chapter 1 -- General Materials

1. Agnew, John A. 1982. "Technology Transfer and Theories of Development." ***Journal of Asian and African Studies*** **17: 16-31.**

Informative article built on the basic premise that "all theories of economic development and social change contain within them more or less implicit positions on the role and impact of technology transfer on development" (p.16). Identifies three groups of theories: modernization (focus on national societies as basic units of analysis, defining development in terms of economic growth and Westernization); world political economy theories (focus on structure of economic and political relations between dominant and dominated countries); ecopolitical economy theories (critical perspective on purported benefits of development emphasizing ecological and social/cultural disruption).

Reviews four perspectives on innovation: (1) process by which adoption of innovation occurs, (2) market and infrastructure context for innovation, (3) continuous nature of adaptation of innovation to particular circumstances, and (4) preconditions for and consequences of innovation diffusion in terms of perpetuating traditional patterns of development and underdevelopment.

2. Ahmad, Aqueil. 1981. "Sociologically Oriented Studies of Science in India." ***International Journal of Contemporary Sociology*** **18: 135-65.**

History of science, science policy studies, and behavioral approaches (management studies) to the study of research are reviewed since the 1960s. Weaknesses of science policy studies suggest a moratorium is needed: the authors of these studies tend to be top scientist-bureaucrats who keep on repeating the same themes. General divorce of research from application. Includes review of institutional structure for science studies as it existed in 1970s. Uncritical and superficial praise of self-reliance in S&T and nuclear R&D is based on Nehru's pronouncements and Indira Gandhi's support. The highly bureaucratized R&D sector doesn't depend on the market and doesn't generate ideas with a chance of use. There is a lack of contact with users. Three kinds of policy-related statements:

factual, normative without theory, and prophetic without theory. Policy studies lack micro focus, only relate to broad trends.

3. Bass, Thomas A. 1990. *Camping With the Prince and Other Tales of Science in Africa.* New York: Penguin.

Recounts the author's experiences on seven different scientific expeditions into Africa, from Timbuktu to the Zambezi River. He camped with Bozo fishermen on the Sahara's edge, dove into Lake Malawi, trapped tsetse flies in the Rift Valley, studied slash-and-burn agriculture in West Africa, "hunted wild viruses" on the Niger and excavated bones and tools on the shores of Lake Idi Amin Dada. Offers an optimistic assessment of the scientific potential in African culture.

4. Bowonder, B. 1981. "Environmental Risk Assessment Issues in the Third World." *Technological Forecasting and Social Change* 19: 99-127.

Written by a policy analyst in Hyderabad, India. Environmental risk assessment is as much a concern to the Third World as to developed countries, but little formal technology assessment exists to carry it out. Environmental degradation in the Third World arises both from development to provide for basic needs and from lack of development and poverty. Major impending risks for the Third World are urbanization, water pollution, bilharzia, damage to soil and agricultural systems, malnutrition and health hazards. Risk assessment involves anticipation of risks and their impacts as well as their assessment and evaluation. Importance of culture in risk perception and anticipation. Deforestation considered as an illustration of these issues.

5. Browett, John. 1985. "The Newly Industrializing Countries and Radical Theories of Development." *World Development* 13: 789-803.

Argues that neo-Marxist dependency theories have failed to account for peripheral capitalist development in the four NICs of East and Southeast Asia, but that this does not necessarily support the "stages" approaches of modernization theory or of Warren's orthodox Marxism. Rather, the phenomenon of uneven development should be understood in terms of the internationalization of capital. Nice summaries of dependency, modernization theory, and orthodox Marxism.

6. Clarke, Robin. 1985. *Science and Technology in World Development.* New York: Oxford University Press.

Originally a UNESCO report developed for its Medium Term plan (1984-9) by an advisory panel on science, technology, and society, this book serves as a concise introduction to the subject, intended for students and the general public, by a British science writer. Many quotes from UNESCO reports. Critical perspective on science with separate chapters on the relationship between science and technology, technology for development, problems associated with development, basic trends in science and technology, the scientific community, the public, the nation state, the transnational corporation, and international cooperative projects.

7. Contractor, Farok J. 1983. "Technology Importation Policies in Developing Countries: Some Implications of Recent Theoretical and Empirical Evidence." *Journal of Developing Areas* 17: 499-520.

Government regulation of technology imports was prominent throughout the world in the 1970s. Intervention seeks to affect three aspects of imports: mode of association between foreign supplier and local operation (full ownership to contractual relation); cost or price of transfer (balance of payments, restraints on local firm); and content of technology transfer package (information, services, rights and restraints). Uses recent data to challenge the basis of LDC importation policies. It is difficult, if not impossible, for governments to effectively regulate the mode, cost, and content of technology imports because of incompatible national objectives.

8. Contractor, Farok J. and T. Sagafi-Nejad. 1981. "International Technology Transfer: Major Issues and Policy Responses." *Journal of International Business Studies* 12: 113-35.

This article is now dated but was a major review (prior to Reddy and Zhao). Covers material from the 1970s, the period when international technology transfer emerged as a field. Covers the nature of tech transfer as a bundle of information, rights, and services; the role of international patents and trademarks; the mode of transfer; costs and compensation; choice of technique; and policy responses in four types of organizations (recipient country, corporate suppliers, supplier government agencies, and the UN). 165 references.

9. Covello, Vincent T. and R. Scott Frey. 1990. "Technology-Based Environmental Health Risks in Developing Nations." ***Technological Forecasting and Social Change*** **37: 159-180.**

Reviews major types of environmental health risks facing developing nations: the failure of large-scale technological systems, the use or misuse of consumer goods, mechanical devices, chemicals, and industrial emissions of toxic substances. Developing nations are particularly vulnerable to technology-related environmental health risks owing to limited risk assessment and management capabilities: inadequate risk data, shortage of risk analysts and facilities, limited financial resources, inadequate environmental legislation, shortage of managers, and limited public awareness and political participation. Export of hazardous products (pesticides, cigarettes, infant formula, therapeutic drugs) and production processes (asbestos, pesticides) as well as hazardous waste from developing nations contribute to this problem. Remedial actions proposed by governments in developed countries, developing nations, international organizations, and MNCs are discussed.

10. Crane, Diana. 1977. "Technological Innovation in Developing Countries: A Review of the Literature." ***Research Policy*** **6: 374-95.**

An interorganizational approach to technological innovation. Inadequate linkages between institutions in various sectors of developing societies inhibit indigenous innovations: few relations between state research organizations and their users. Academic research is (mis)oriented towards the international scientific community. Coordinating agencies have also performed poorly because of lack of influence with other organizations. Bibliography classified by region.

11. Dahlman, Carl J., Bruce Ross-Larson and Larry Westphal. 1987. "Managing Technological Development: Lessons from the Newly Industrializing Countries." ***World Development*** **15: 759-75.**

Uses the case description of an archetypically successful firm for introducing main concepts. Acquiring the capabilities for efficient production and investment is more important for successful industrialization than the invention of products/processes. The drive for self-sufficiency can result in poor productivity: foreign and local technological elements must be combined. Technological capability does not simply result from experience, but from monitoring worldwide. The economic environment is composed of penalties and incentives affecting the way firms use and adapt technology.

12. Erber, Fabio Stefano. 1981. "Science and Technology Policy in Brazil: A Review of the Literature." ***Latin American Research Review*** **16: 3-56.**

Review of texts treating the relationship between socio-economic development and the development of S&T. First part discusses theories of the role of S&T in capitalist development (e.g., aggregate production functions, historical approaches). Until recently, acceleration of the rate of innovation was accepted as a fact, but innovations are not a linear function of R&D expenditures. Second part discusses specific conditions in Brazil--doubts in the 1960s about benefits arising from the international division of technical labor, suggestions that measures be taken to develop technical and scientific capacity. Reviews studies by IPEA on technological research conducted in Brazil and tech transfer from abroad, technological dependency studies, studies of the scientific and technological system.

13. Evans, Donald D. and Laurie Nogg Adler (eds.). 1979. ***Appropriate Technology for Development: A Discussion and Case Histories.*** **Boulder, CO: Westview Press.**

Contains case studies of specific appropriate technologies in 22 different countries. The purpose is to present an overview of contemporary thought on the relationship between technology and the economic development of less-successful nations and to present detailed information on the application of technology in locations throughout the world. Most of the cases were written by members of the Denver Research Institute. Brief abstracts of the 22 studies are given at the beginning of the book. Offers a system for arraying the case histories according to a structure of interactions and effects. It is held that cases may be examined in terms of three primary elements: (1) human and material resources with which they are concerned, (2) the means by which the resources are developed or utilized, and (3) the effect that utilization has on the environment and people. The effects may be classified as economic effect, social impact, and the effect on the physical environment. Concludes with a good bibliography of works on AT published in the 1970s. Each entry has a short annotation and key terms.

14. Evans, Peter and John Stephens. 1988. "Development and the World Economy." Pp. 739-73 in *Handbook of Sociology*, edited by Neil Smelser. Newbury Park, CA: Sage.

Excellent review essay, including empirical work, arguing that modernization and dependency perspectives have been synthesized. Although science and technology are not an emphasis, this is essential reading.

15. Forje, John W. 1982. *Science and Technology for Development in Africa South of the Sahara*. Lund, Sweden: Research Policy Institute.

A bibliography of materials on S&T for development in Africa. All works listed are from the 1970s, with most from the mid to late 70s. Contains 649 alphabetical listings (by author or organization) and 24 periodicals. Offers a subject index and a regional index, broken down into Africa General, Central Africa, East Africa and the Indian Ocean, Southern Africa, and West Africa.

16. Forje, John W. 1989. *Science and Technology in Africa*. London and New York: Longman.

Part of the Longman Guide to World Science and Technology series. An overview of S&T in Africa. Examines the geography and population of Africa and divides the continent into seven main subregions (North Africa, Sahel, Non-Sahelian West Africa, Central Africa, East Africa, Southern Africa, and islands of the Indian Ocean). Discusses the historical development of African S&T. Looks at science and government, national science and technology policy, and government structures and mechanisms for S&T policy. Considers human resource issues such as education, institutions, scientists and engineers, etc.

Contains chapters devoted specifically to agricultural S&T, medical sciences and biotechnology, industrial R&D, R&D in the petroleum industry, energy and development, and computer technology and development. Discusses cooperation in S&T at the international, regional, and subregional levels, as well as South-South cooperation. Discusses the role of military buildup in African development. Describes means of popularization of S&T, such as the media, professional societies, and libraries and computer technology. Contains a chapter on transport and communications. Appendices contain the

Brazzaville Declaration on Science and Technology, the Kilimanjaro Declaration, and a directory of research establishments and organizations.

17. Fransman, Martin. 1985. "Conceptualising Technical Change in the Third World in the 1980s: An Interpretive Survey." ***Journal of Development Studies*** **21: 572-652.**

Excellent, in-depth review essay by University of Edinburgh economist focusing on developmental economics in the late 1970s and early 1980s. About 250 references. Concerned with technology in the manufacturing sector. Covers changing conceptualization of technical change (technology market, transfer of technology, choice of technique), its sources and measurement. Sections are devoted to the role of the capital goods sector, analysis of infant industries and exporting activities, the role of the state, and the question of whether developed country technology is an opportunity or constraint. Concentrates on the more industrialized countries of the Third World with some contrasts with Japan.

18. Furtado, Celso. 1977. "Development." ***International Social Science Journal*** **29: 628-50.**

Part of an issue on interdisciplinary social science. Historical view of the development of development. Sees both development and underdevelopment as having their roots in the acceleration of accumulation in Europe at the end of the nineteenth and beginning of the twentieth century, which led to the world economic system of industrial capitalism. Devotes sections to the dissemination of instrumental rationality and to the role of technology in the reproduction of capitalist society.

19. Gareau, Frederick H. 1985. "The Multinational Version of Social Science with Emphasis upon the Discipline of Sociology." ***Current Sociology*** **33: 1-165.**

Sociology of the social sciences. Concept of social scientific "sects" may be applicable to the hard sciences as well. Examines social scientists' pursuit of John Stuart Mill's dream of applying the methods of the physical sciences to the social sciences. In Kuhnian terms this would transform social scientists into problem solvers. Social science is elitist and oligarchic--most of those cited in prestigious journals received degrees from a small number of prestigious universities. Useful concepts for the study of science in Third World development--what is the pattern of relations between scientists, where do high-status scientists get their degrees, what are the personal and professional ties among development scientists?

20. Ghosh, Pradip (ed.). 1984. *Technology Policy and Development: A Third World Perspective.* Westport, CT: Greenwood Press.

Number 3 in the series International Development Resource Books. Concerned mainly with technology gaps between developed and developing countries, technology changes in developing countries, and progress toward self-reliance in developing countries. Part I contains articles on these issues, many reprinted from Impact of Science on Society (UNESCO journal). Part II through Part IV contains statistical tables and figures, bibliographies, and directories, sections that are similar to the corresponding sections in Ghosh and Morrison's book *Appropriate Technology in Third World Development.*

21. Ghosh, Pradip and Denton E. Morrison (eds.). 1984. *Appropriate Technology in Third World Development.* Westport, CT: Greenwood Press.

Number 14 in the series International Development Resource Books. Reprints articles by a variety of authors in order (1) to document and analyze AT policy trends and (2) to evaluate progress in AT made by Third World countries over the previous decade. Part I concerns current issues, trends, analytical methods, strategies and policies, and country studies. Parts II-IV are of particular value to researchers. Part II, on statistical information and sources, contains a bibliography of information sources, as well as statistical tables and figures. Part III, a resource bibliography, contains a well-annotated bibliography of books, focusing mainly on development and AT (most sources pre-1980). A listing is provided, with occasional brief comments, of selected periodical articles, and abstracts of empirically relevant studies. Part IV, a directory of information sources, gives addresses of UN information sources, a bibliography of bibliographies, a directory of periodicals, and addresses of research institutions.

22. Goss, Kevin F. 1979. "Consequences of Diffusion of Innovations." *Rural Sociology* 44: 754-72.

Review of diffusion of innovation literature with specific reference to developing countries. Case studies of Colombia (institutional constraints) and Bangladesh/Pakistan (structural differentiation). Future research should focus on macro-level variables (especially distributional) rather than social-psychological variables and the consequences of diffusion as dependent variables.

23. Goulet, Denis. 1989. *The Uncertain Promise: Value Conflict in Technology Transfer*. New York: New Horizons.

After confronting the "two-edged sword" represented by the dynamics of the "technological universe," the author evaluates whether technology transfer is an aid or an obstacle to development, focusing on channels of transfer, the role of TNCs and the high price paid by LDCs (including numerous Latin American case studies). Review of development strategies and Third World technology policies within the context of the world order. Concludes with a philosophical discussion of the issues involved.

24. Gupta, Avijit. 1988. *Ecology and Development in the Third World*. London: Routledge.

Brief, introductory volume covering developmental effects on natural vegetation, water resources, air quality, and other environmental variables, with ten brief case studies.

25. Heller, Peter B. 1985. *Technology Transfer and Human Values: Concepts, Applications, Cases*. Lanham, MD: University Press of America.

Twenty-one case studies of technology transfer by agents from the World Bank, the Peace Corps and other agencies dealing with appropriate technologies and the myriad changes that occur as the result of transfer. A general introduction covering transfer and diffusion, human values, alternative and inappropriate technologies, barriers in transfer, and public policy. Concise summaries of the case studies, keyed to the introductory chapters, and a glossary make this a good volume for teaching purposes.

26. Hollick, Malcolm. 1982. "The Appropriate Technology Movement and Its Literature: A Retrospect." *Technology In Society* 4: 213-29.

Criticizes the Appropriate Technology movement as biased by Western, individualistic, middle-class pastoral myths that hide the realities of Third World life. "Small is not always beautiful." Rural self-sufficiency can be painful, and not all developing economies can utilize craft production. Questions the desirability of always promoting cultural stability at the expense of technology.

27. Hurley, D. 1987. *The Management of Technological Change: An Annotated Bibliography*. London: I.T. Publications and Commonwealth Secretariat.

Jointly prepared by the Intermediate Technology Development Group and the Commonwealth Secretariat, this bibliography presents abstracts of articles and books by over 140 authors, including case study information on over 49 developing countries. Most of the publications are from the late 1970s and early 1980s. These are arranged in six chapters: (1) Technology and Development, (2) The Policy Framework, (3) International Technology Transfer, (4) Technological Change within Developing Countries, and (5) Bibliographies.

28. Inkster, Ian. 1991. "Made in America But Lost to Japan: Science, Technology, and Economic Performance in the Two Capitalist Superpowers." *Social Studies of Science* 21: 157-78.

Excellent essay review of books by Giovanni Dosi's group and an MIT group on national economic development and technology. Says quite a bit about contemporary economic views.

29. Ives, Jane H. 1985. "Health Effects of the Transfer of Technology." Pp. 172-192 in *The Export of Hazard: Transnational Corporations and Environmental Control Issues*, edited by Jane H. Ives. Boston: Routledge and Kegan Paul.

An overview of the health effects of technology transfer and a review of recent case studies, presenting current international debates and regulatory issues. Recommends the development of "... a clear, consistent and rigorous policy within the World Health Organization." Gives a good bibliography on this topic and a list of information resource and research centers on the health effects of transfer of technology to the developing world.

30. James, Dilmus D. 1988. "Accumulation and Utilization of Internal Technological Capabilities in the Third World." *Journal of Economic Issues* 22: 339-53.

Presidential Address for the Association for Evolutionary Economics but also a highly readable review article. The 1980s were characterized by more self-reliance of LDCs as firm-level innovation studies, and country-specific investigations of technology exports showed that upper-tier LDCs can achieve local mastery over technology. Review of case studies and policy options. LDCs should narrow choice in R&D to fewer projects, shift funds to applied work, avoid projects with international prestige but removed from problems.

31. James, Jeffrey. 1978. "Growth, Technology and the Environment in Less Developed Countries: A Survey." *World Development* 6: 937-66.

Shows interrelationship between growth, technology and changes in the environment, but doesn't find direct evidence that environmental problems are caused by "careless technology." The first section looks at general approaches to environmental problems in the literature, including physical laws governing the relationship between economic activity and the environment, the welfare economics approach (concerned with the rapidity of depletion of non-renewable resources on the output side and the limited capacity of the environment to assimilate residuals from processes of production and consumption on the input side), and the technological approach (whose leading exponent is Barry Commoner, arguing that damage to the environment stems from technology choices rather than economic activity per se).

Suggests that one difficulty with the technological approach may be the inseparable mutual causality of technology, population, and affluence. Investigates the concept of the steady-state economy as an implication of the technological approach. The second section compares LDCs with developed countries to describe the nature and causes of environmental disruption in the Third World. Sections 3 and 4 discuss aspects of international environmental interdependence.

32. Jamison, Andrew. 1988. "Innovation Theories and Science and Technology Policy: Historical Perspectives." Pp. 17-53 in *Science, Technology, and Development*, edited by Atul Wad. Boulder: Westview Press.

Looks at the historical development of innovation theories in relation to science and technology policies. Suggests that there has always been interdependence between policy and theory. Attempts to indicate how policy doctrines and academic discourse can be seen as responses to change in scientific/technological practice.

Discusses the precursors of innovation theories in the 16th through the 18th centuries. By the late 18th and early 19th centuries, Jamison sees two approaches among theorists of innovation: "external" approaches, such as Adam Smith's, that focus on technology in the economy and social system; and "internal" approaches, such as Diderot's, that focus on technology in its own right. During the 19th century, these two approaches were combined and technological development was viewed largely as a political issue.

Marx was one of the first to try to grasp the transformation of innovation in terms of its scientific/technical dimension. Marx (1) drew a fundamental distinction between tools and machines, (2) placed the technological innovation process into a comprehensive historical scheme, and (3) drew attention to the new kind of relationship emerging between science and technology. Discusses development of Marxist views on these issues and looks at a variety of early 20th century views of technology, including Veblen, Mumford and Ogburn. After WWII, views of S&T policy are described as based on the assumption of "science-push." In the 60s and 70s, policy was seen more in terms of social assessment. By the end of the 70s, there was a shift to "the new economic context." The notes give good bibliographical information on S&T policy.

33. Johnston, Ann and Albert Sasson (eds.). 1986. ***New Technologies and Development.*** **Paris: UNESCO.**

Papers from a UNESCO colloquium on science and technology as factors of change (Paris, 1984). The chapters consider information technologies, biotechnologies in farming and food systems, health care technologies, energy technologies, and technological advance and social change. Takes the general view that cooperative research activities between Third World countries and industrialized countries or multinational organizations will be necessary for a long time, but that cooperation among developing countries will become increasingly important. The Selective Bibliography in the back of the book will be particularly useful for research. Areas covered: general, Club of Rome publications, prospective studies in science and technology, regional studies, energy, employment and technological change, and methodology. Works in English, French, and German.

34. Lall, Sanjaya. 1984. "Exports of Technology by Newly-Industrializing Countries: An Overview." *World Development* 12(5/6): 471-480.

This issue of *World Development* focuses on exports of technology by newly industrializing countries. It includes case studies of Hong Kong, Taiwan, the Republic of Korea, India, Egypt, Brazil, Mexico, Argentina, and Latin America as a whole. Lall's overview contains a literature review of the analysis of Third World technology exports, a summary overview of technology exports, and a report on the indigenous technological content of technology exports.

35. Lall, Sanjaya. 1992. "Technological Capabilities and Industrialization." *World Development* 20: 165-86.

Reviews implications for industrial strategy of recent research on technological capabilities at the firm and national levels. Sets out a framework for explaining the growth of national capabilities, based on the interplay of incentives, capabilities, and institutions. Describes the experience of some industrializing countries in order to assess the validity of the framework. Concludes that careful and selective interventions are needed for industrial success.

36. Latin American Newsletters Limited. 1983. *Science and Technology in Latin America*. London and New York: Longman.

Volume 2 in the Longman Guide to World Science and Technology series. A country-by-country examination of current state of S&T in Latin America. The geographical, demographic, economic, and political features of each country are given (except in the case of Cuba, where the political aspect is largely omitted). This is followed by descriptions of the organization of science and technology in each country, with flow charts showing relations of governmental and scientific organizations. Policies and financing in science and technology are then discussed.

Separate sections are devoted to science and technology at the governmental and academic levels. Both of these take the form of lists of institutions and organizations, with brief mentions of their functions.

This is followed by lists of organizations involved in specific scientific and technological fields. Appendix 1 gives regional statistics, including socio-economic indicators, education, national expenditures on R&D, numbers of individuals employed in R&D activities. Appendix 2 is a directory of selected establishments in each country, with addresses.

37. Longman Guides, various. *Longman Guide to World Science and Technology*. Detroit: Gale Research Co.

A series with in-depth coverage of scientific and technological conditions for world regions. Pertinent books include coverage of science and technology in: *Science and Technology in Africa*, by John W. Forje (#10, 1989), ...*Australasia, Antarctica and the Pacific Islands*, by Jorlath Ranayne and Campbell Boag (#11, 1989), ...*China*, by Tong B. Tang (#3, 1984), ...*the Indian Subcontinent*, by Abdur Rahman (#13, 1989), ...*the Middle East*, by Ziauddin Sardar (#1, 1982), and ...*Latin America*, by Latin American Newsletters Ltd. (#2, 1983).

38. MacCormack, Carol P. 1989. "Technology and Women's Health in Developing Countries." *International Journal of Health Services* 19: 681-92.

A review article, concerned with access to and control of medical technologies by women in developing countries. Seeks to assess whether the literature is adequate to guide planners wishing to enhance the health of Third World women and whether the conceptual models being used to interpret reports and studies are consistent with the aims of efficient use of resources and social justice.

39. Milic, Vojin. 1990. "Recent Trends in Science and Technology in Developing Countries." *Sociology* 32: 351-400.

Discussion of the development and state of science organization in developing countries. Broad socio-cultural regions of development are described (Arab countries, Latin America, Asia, Balkans, Sub-Saharan Africa, Iberian Peninsula and Ireland) after a discussion of conceptual deficiencies of the term "developing countries." Founding of autonomous states is a precondition for indigenous science, including the founding of universities. Summary of data on growth rates, illiteracy, founding of universities, and R&D. Ends with discussion of the importance of language indigenization in integrating science into a culture.

40. Rath, Amitov S. 1990. "Science, Technology, and Policy in the Periphery: A Perspective From the Centre." ***World Development*** **18: 1429-1445.**

Introducing a special issue on "Science and Technology: Issues from the Periphery." Articles cover theoretical underpinnings, Latin American manufacturing, technology policy in Sub-Saharan Africa, science and technology indicators, and environmental degradation in peasant agriculture. This leading article discusses research at the International Development Research Centre in Canada. Four broad areas of S&T policy concern development: international context, the formal R&D system, the production system, and the impact on society.

41. Reddy, N. Mohan and Liming Zhao. 1990. "International Technology Transfer: A Review." ***Research Policy*** **19: 285-307.**

Excellent review of technology transfer studies between countries, organized around supplier country, recipient country, and transfer itself. Economic studies dominate. Summary includes directions for future research and laments that ITT studies have not tapped the technology diffusion literature. Topics covered are economic benefits and costs to home and host countries, government policy, MNCs, adaptations to host country, LDC regulatory approaches.

42. Robinson, R.D. 1991. ***International Communication of Technology: A Book of Readings.*** **London: Taylor and Francis.**

Part of International Business and Trade series, containing articles on variables related to international technology transfer, multinational enterprises, technology transfer to China, and transfer of organizational technology.

43. Rybczinski, Witold. 1991. ***Paper Heroes: Appropriate Technology: Panacea or Pipe Dream?*** **Garden City, NY: Anchor Press/Doubleday.**

Republished version of 1980 volume which reviews basic texts of the appropriate technology movement, offers a critique, and signs the death certificate. Readable text with photos.

44. Sardar, Ziauddin. 1982. ***Science and Technology in the Middle***

***East*. London and New York: Longman.**

Volume 1 of the Longman Guide to World Science and Technology series. Part I is an overview of S&T in the Middle East, covering science planning and organization, scientific manpower and the brain drain, science education and Arabization, Islamic science, medicine and health, agriculture and irrigation, environment and pollution, nuclear energy, solar energy, microtechnology and communications, information science and international centers, regional cooperation, scientific and technical assistance, future trends, and a fairly extensive bibliography of books and articles on S&T in the Middle East.

Part II describes regional organizations concerned with science and technology. Part III offers country profiles that give the demographic and economic background of 19 Islamic countries and describe S&T institutions and current research in each. These 19 countries include the Maghreb (Morocco, Tunisia, and Algeria) and Islamic South Asia (Afghanistan and Pakistan), as well as countries that would be included in a narrower definition of the Middle East.

45. Shrum, Wesley and Yehouda Shenhav. 1994. "Science and Technology in Less Developed Countries." In *Handbook of Science, Technology, and Society*, edited by James Petersen, Gerald Markle, Sheila Jasanoff, and Trevor Pinch. Newbury Park, CA: Sage.

Review essay covering the social science literature from the mid-1970s to the present, with an emphasis on connections to current social studies of science and technology.

46. Simmons, John and Leigh Alexander. 1978. "The Determinants of School Achievement in Developing Countries: A Review of the Research." *Economic Development and Cultural Change* 26: 341-58.

Determinants of achievement are largely similar to those in developed countries. See also the reply in volume 28, number 2 (Jan. 1980).

47. Sussman, Gerald and John A. Lent. 1991. "Critical Perspectives on Communication and Third World Development." Pp. 1-26 in *Transnational Communications: Wiring the Third World*, edited by Gerald Sussman and John A. Lent. Newbury Park, CA: Sage Publications.

Review of literature on Third World communication systems. Offers a very critical overview of the modernization perspective, represented by Daniel Lerner, Ithiel de Sola Pool, Wilbur Schramm, and Lucien Pye. The authors favor a "grass roots approach," developed by Freire, Schumacher, Diaz Bordenave, and Fugelsang. They are sympathetic to dependency critiques and to the demands of non-aligned nations for a New International Economic Order (NIEO) and a New World Information and Communication Order (NWICO) during the 1970s. Communication literature during the 1980s was dominated by postindustrial theses of Daniel Bell, Alvin Toffler, Wilson Dizard, and Everett Rogers, who are criticized as being technological determinists. Concludes with an examination of the political economic perspective, which includes dependency theorists, Marxian critics of dependency, world systems theorists, and others generally opposed to transnational corporations and the dominant position of the West.

48. Tang, Tong B. 1984. *Science and Technology in China*. London and New York: Longman.

Volume 3 in the Longman Guide to World Science and Technology series. The introduction gives a brief overview of the demographic and economic background of China, China's science policy, science organization, tertiary education, human resources and science popularization, and international exchanges and information management. This is followed by a list of publishing houses for science and technology in China. Other chapters deal with science and social change in China, biomedical research, agriculture and environmental research, earth sciences, mineral industries, energy science and technology, transportation and information transformation, and science and technology in Taiwan.

Appendix 1 contains statistical tables on human and natural resources, administrative division and population distribution, agricultural production, primary energy reserves estimates, recoverable reserves and output in 1982, crude coal production, and crude oil production and consumption. Appendices 2-6 describe each of China's five-year plans. Appendix 7 gives a summary of the 1978-85 science plan, taken from Vice-Premier Fang Yi's speech at the All-China Science Conference in March, 1978. Appendix 8 give a directory of major research establishments (without mailing addresses).

49. Theofanides, Stavros. 1988. "The Metamorphosis of Development Economics." ***World Development*** **16: 1455-63.**

A historical account of the development of Development Economics, from a paradigm to thirty specialties, arguing against those who have pronounced Development Economics dead. Specialization is an indicator of progress; the discipline has "desynthesized" and shifted away from pure economics to non-economic subjects. Lists the subdisciplines and the journals.

50. Vernon, Raymond. 1990. "Trade and Technology in the Developing Countries." Pp. 255-270 in ***Science and Technology: Lessons for Development Policy*****, edited by Robert E. Evenson and Gustav Ranis. Boulder: Westview Press.**

Review of studies dealing with the relation of technology to international trade for developing countries. Attempts to examine the central propositions that have repeatedly emerged from these studies. Topics include the effect of the inefficient market for technology on the transfer of technology, choice of foreign channels for technology transfer (covering preferred policies and characteristics of wholly owned subsidiaries, joint ventures, and licensing agreements), attempts to categorize industrial technologies, and the role of government (dealing with market failure and gauging the effect of government involvement).

51. Welsh, Brian W.W. and Pavel Butorin. 1990. ***Dictionary of Development: Third World Economy, Environment, Society*****. Hamden, CT: Garland Press.**

Two-volume guide to international organizations, institutions, programs, and conferences involved in development issues. The profiles of organizations include their address, goals, purpose, projects, budgets.

52. Yearley, Steven. 1988. ***Science, Technology, and Social Change.*** **London: Unwin Hyman.**

The work as a whole stresses technology and development in the context of recent work in social studies of science and technology. Focuses on development and modernization theory in chapter 6. The only work that we know of providing this overview of new developments in social constructivism.

Part II -- The Exchange and Diffusion of Innovations: A Review of Major Themes in the Literature

The literature on science and technology has emerged from a wide variety of disciplines and perspectives. Its concerns and subjects are equally varied. Despite the apparent diversity, however, an underlying unity runs through much of this literature in the form of related recurring issues tied to a few basic orientations. This essay will attempt to delineate these issues in a schematic fashion.

A broad topic that runs through current writing on science and technology may be referred to as the exchange and diffusion of technology. This involves all of the concerns that cluster around the movement of objects, ideas, or forms of social relations between individuals, groups, and places. "Exchange" refers to interactions at various levels. "Diffusion" refers to the spread of the objects, ideas, and behaviors involved in exchanges.

Two predominant components of the spread of technology are the ideas of technology transfer and innovation diffusion. Technology transfer usually is treated as the transfer of artifacts and skills across national boundaries, although it also may be seen as transfer between domestic organizations or sectors, as in Baark's study of the transfer of technology from research to production units in China (123). A count of the pieces in this volume shows that at least one-fourth of the writing on science and technology in developing countries is explicitly and directly concerned with the subject of technology transfer.

Innovation diffusion is conceptually connected to technology transfer, although Cottrill et al. (478) have suggested that these two subjects remain tied to two separate academic literatures. Since many authors consider the effectiveness of technology transfer to be a function of the ability of countries to absorb and utilize new technologies, the close logical association of the two subjects is evident (cf. Heller 25; Stewart and Nihei 247; Cole and Mogab 143; Mytelka 210; Chepaitis 140; Colle 364; Ali 120).

Because most developing country technologies are acquired through international transfer, rather than through domestic innovation (Pack and Westphal 332), these topics may, in fact, be regarded as inseparable. There is, however, a substantial literature stressing the extent to which domestic innovation is preferable and possible (Hainsworth 810; Herrera 381; Mohan 204; Teitel 409; Zahlan 266), as well as a lively debate over whether technology transfer encourages or inhibits the local production of innovative techniques (Segal 228; Blumenthal 359; Katrak 182; Kumar 187; Siddarthan 234; Al-Ali 117; Dahlman and Westphal 147).

Issues in Technology Transfer

Economists interested in technology transfer frequently study how and to what extent transfer affects the development process. At the most abstract level, this involves constructing models and theories of economic development, as in the work of Lee (511) or Bruton (134). Agnew (1) has suggested that all theoretical perspectives, whether economic or social, contain implicit positions on the role and impact of technology transfer on development.

A theory of the relation of technology transfer to economic development also seems to imply views on the best strategies for using transfers to promote development. Technology strategies may be seen as consisting of both technology choice and choice of transfer strategy. Agnew's suggestion (1) offers the key to constructing a framework for the literature in this area: a basic element of the relevant theoretical perspectives is how these perspectives see supplier-recipient relations. Some authors organize the technology transfer literature in terms of the concepts of supplier country, recipient country, and the transfer itself (Reddy and Zhao 41).

Those who take supplier-recipient relations as essentially and systematically unequal usually concentrate on the issue of inequality itself and how this constrains or renders ineffectual the decisions of developing countries (cf. Stewart 101; Stewart and James 346; Girvan 161; Yapa 413; Vayrynen 256; Servaes 230; Long and Pollard 518; Kaplinsky 177; Al-Ali 117; Beaumont et al. 128). Some have seen the problem of unequal relations as rooted in the values of developing and developed, as well as in their economic positions. These authors have suggested that the transfer of Western technology to Less-Developed Countries (LDCs) is also the transfer of Western values, values that are inconsistent with the needs and development objectives of LDCs (Bailey et al. 563; Riedijk 93; Servaes 230). Those who see technology transfer as problematic, but unavoidable, often portray choice and strategy as a delicate balancing act between benefits and costs (Stewart 245; Stewart 246; Hamelink 76). From the point of view of developing countries, some have suggested that the best answers to unequal relations may lie in increased South-South trade (Hamelink 165; Sagasti 665; Sharif and Haq 232) and in efforts to achieve greater independence (Stewart and James 346; Mohan 204). Several scholars have attempted to suggest ways of restructuring the international system itself, such as a code to govern North-South transfers (Ramesh and Weiss 217; McCulloch 637).

Those who believe that the current structure of supplier-recipient relations, although unequal, can actually be beneficial to developing countries tend to emphasize the decisions or steps developing countries and supplier countries (as well as organizations) should take. This often involves empirical studies of how international technology transfer contributes to domestic progress. Frequently, such studies take neo-classical economic perspectives, as in Havrylyshyn's paper (305), which argues that South-South trade is economically less efficient than North- South trade. Authors who take this perspective have presented empirical evidence that foreign investment can contribute both to the scientific development of LDCs (Lipsey 516; Blumenthal 359) and to their employment levels (Koizumi 314; Chen 139). Baranson (125) and Baranson and Roark (126) have suggested that the international system of exchange relationships is gradually becoming more advantageous for developing countries, pointing to increased technology sharing on the part of U.S. corporations, to the shift toward high technology among the Newly Industrialized Countries (NICs), and to the upward movement in the international division of labor.

A number of studies explicitly recognize the conflicts of interest that exist among suppliers, but insist that reaching compromises on these conflicts is possible, to the mutual advantage of both parties. Thus, Georgantzas (160) has offered a "cognitive mapping model," suggesting that bargaining between LDC representatives and Multinational Corporation (MNC) representatives can lead to mutually beneficial results. According to Madu and Jacob (196), disagreements between LDCs and MNCs can lead to effective compromise as long as the two agree on basic issues.

Taking the present state of supplier-recipient relations as given, technology choice and strategies may be seen from the point of view of either the supplier or the recipient. Occasionally, those concerned with supplier choices may be interested in how the North can best contribute to the progress of the South (as in Morgan 207). More often, however, since the ultimate benefits of the exchange have been demonstrated or assumed, those who examine supplier choice are concerned with the question of profitability. This may involve questions such as how suppliers can protect their interests through effective grants of property rights (Burstein 279) or how they can achieve efficient organizational structures (Davidson 149). Although competition usually is assumed to be fundamental to the behavior of supplier organizations, dealings between competing firms are dealt with only rarely (as in the highly technical article by Marjit 199).

Michalet, and those influenced by him, have examined how profitability has determined the technology transfer strategies of multinational corporations, finding that R&D has become more centralized in MNCs, and that MNCs have moved from stressing patents and licenses to a strategy of dissemination in order to prevent leaks of technological elements (Michalet 203; see also Wionczek 263). Related to this outlook on the institutional results of the search for profitability by suppliers is the "appropriability theory" of MNCs. This theory argues that suppliers of technological innovations expand in order to internalize the advantages of new information (Magee 324).

Technology transfer is much more frequently seen from the point of view of the recipient, contrary to the findings of Cottrill et al. (478), who concluded that the technology transfer literature usually is concerned with suppliers while the literature on innovation diffusion usually is concerned with users. This may be a result of the fact that the sources used in Cottrill's co-citation analysis are somewhat dated, drawn from the period 1966-1972.

Scholars who approach the problem from the recipient perspective tend to be concerned with the benefits developing countries can derive from technology transfer and how to maximize these benefits. Samli (225) has maintained that technology transfer is the most viable way for developing countries to break out of a "vicious cycle" of underdevelopment. Weiss (349) has attempted to describe the contribution of a market-oriented S&T policy as a series of stages to modernization, involving technology transfer during a process of achieving technological mastery. D'Ambrosio (726), also, sees technology transfer as a necessary first stage in development. Derakshani (151) has examined the factors affecting the successful transfer of technology.

Benefits may be maximized by making wise choices regarding types of technology to be imported or by devising an effective S&T policy. Regarding technology choice, Salas Capriles (224) takes an unusually sanguine approach, assuming that LDCs are free to choose the best technology for their situations autonomously. Others, such as Derakshani (151), do not impute this kind of freedom to the decision makers of recipient countries, but still see decisions on technology transfer as critical to development. M.R. Bhagavan (569) has attempted to delineate S&T strategies that can maximize recipient benefits.

The question of maximization of benefits leads to the issue of appropriate technology. Both those who see supplier-recipient relations as essentially harmful to the interests of developing countries and those who see those relations as potentially or actually beneficial examine the "appropriateness" of imported technologies. However, "appropriateness" is taken in at least two different senses. Those concerned with inequality and negative effects in the transfer process take it to mean "appropriate to the environment and social order of the developing country" (examples are Donaldson 67; Gritzner 163; Hainsworth 810; Hamelink 76, Hungwe 80; Kaniki 84; Lien 762; Riedijk 93; Shaw 836; Yapa 114). Those less critical of the relations in the transfer process tend to concentrate more on the economic appropriateness of technology (improving the developing country's competitive position and corresponding to the country's current stage of development). They may also disagree with the position that "low tech" or "soft tech" is actually appropriate (cf. DeGregori 960; Onn 87; Thoburn 106; Trak and MacKenzie 107; Young 265). Ideological positions, of course, overlap somewhat. Works that examine the appropriateness of particular types of technology or that offer microlevel descriptions of technology in a given country often can coexist with either view of the transfer process (as in Fleissner 71; Greeley 435; Heller 25; Perrolle 447; Riskin 94; Smil 100). "Appropriateness" is a subjective concept, lacking a generally accepted definition (Bowonder 422); evaluating the appropriateness of a given technology is inherently problematic (Sharif and Sundararajan 677).

Just as scholars disagree on what constitutes an appropriate item of technology, they also disagree on what constitutes an appropriate technology policy. Some, such as Mohan (204), seem to oppose all forms of technology transfer, and therefore would seem to advocate highly protectionist policies aimed at promoting indigenous innovation. Tho (347), among many others, has advocated a policy of substituting local resources for foreign capital. Clarry (142) views technological dependency as a problem, but one that can be moderated when host country elites set strategic policies for bargaining with MNCs.

Those who hold a more optimistic view of international exchange offer evidence to defend the practice of technology transfer. For example, Desai (289) finds that India needs to pay less attention to the regulation of technology and more to encouraging technology's diffusion. Shishido (680) has defended a technology import policy by pointing out that Japan initially encouraged the import of technology and that this played a significant role in the development of the Japanese economy. Contractor (7) has offered empirical evidence that government regulation of technology imports may be ineffective.

Scholars who do not see the exchange relationship itself as the primary problem often concentrate on the developing country infrastructure as a source of difficulties. For Rosenberg (337), the critical issue is how to link applied science with improved technologies. The issue of intellectual property rights has been raised by those who see these rights as an essential part of a country's S&T infrastructure (see, for example, Evenson 600; Sherwood 679).

Perhaps understandably, those who write on effective government policies, from whatever standpoint, tend to advocate a more or less activist state, even if they eschew protectionism. Successful technology transfer and successful economic development both have resulted at times from government planning and not simply from market forces (Pack and Westphal 332; Tuma 254). Girvan and Marcelle (377) maintain that an active government strategy is essential to effective transfer.

The social effects of the exchange are most often of interest to those critical of the exchange relationship, possibly because others simply assume benefits of the exchange at all levels. However, one piece on the positive social effects of transferred technology, a paper by Von Braun (457), finds that technological change brought higher incomes and higher caloric intake to West Africa. Blyn (421) also provides evidence of the social benefits of technology transfer, finding that tractorization provided for social betterment in Punjab and Haryana.

The undesirable social influence of the Green Revolution is a major theme among critical authors. Many have maintained that Green Revolution technologies increased economic disparity by benefitting wealthy farmers and large landowners at the expense of the poor (for example, Bowonder 422; Burke 423; Chadney 802). Zarkovic (464) has found, also, that the Green Revolution techniques were damaging to the interests of women, since they increased male predominance in the labor force.

Others critical of the social effects tend to fall into two camps. Some, like Servaes (230), argue that Western technology per se is harmful to developing country societies, since it embodies Western values intrinsically destructive of traditional cultures (see also Yapa 413, 114). By contrast, some have argued that greater impoverishment and social and economic disparity does not come from the technology itself, but from the misuse of the technology by an already unequal and unjust society. Ridler (450) has found that increased unemployment and land consolidation are not the effects of technology alone, but of the technology in the developing country setting. Along these same lines, Braverman and Stiglitz (360) have offered evidence that landlords tend to suppress innovations that would lighten the debt loads of their tenants, while they often adopt those that will increase the loads.

In summary, technology transfer is best seen as a form of exchange. The views and interests of those who study this topic are often determined by how they see the exchange relationship. Even authors who have no overt concerns with international relations or political systems tend to be guided by the extent to which they believe this exchange can be beneficial to recipients, and by the positioning and bargaining strategies they believe recipients should adopt to maximize benefits. The literature on science and technology in developing countries is concerned with a larger process, as well as a specific exchange. This process is the way in which the objects, ideas, and behaviors that are exchanged spread in a geographic area or social setting, normally referred to as "innovation diffusion."

Issues in Innovation Diffusion

The adoption of technological innovations spreads through three primary mechanisms: (1) research and development organizations (cf. Biggs and Clay 358; Compton 365), (2) formal education (cf. D'Ambrosio 726), and (3) informal education, including institutional extension agencies and patterns of communication among users (Aina 353; Eastman 370; James 438). Distinguishing among these three mechanisms is to some extent heuristic, since all three coexist and overlap. Research and education are frequently both located in a university setting, and may often be a part of a single process. Informal patterns of communication may be found between members of formal organizations.

Diffusion may be considered on at least two levels. At a macro level, a scholar may be interested in how new technology spreads among countries, firms, or other collective entities (as in Soete 242; Lall 510). At a micro level, the point of interest may be the adoption of an innovation by individual users, or communication between individual users or researchers (as, for example, in Agarwal 350; Colle 364; Compton 365, and Gamser 375). Studies that look explicitly at innovation diffusion tend to concentrate much more on intranational diffusion among users (Goss 22), a fact that might explain the finding of Cottrill et al. (478) that technology transfer and innovation diffusion, although logically connected, remain two separate literatures.

This difference in concentration may account for the tendency to underemphasize the exchange relationship when discussing the diffusion of technology. The supplier-recipient issue arises most often in studies of unsuccessful diffusion. The unequal relationship may be found to have resulted in the attempt to spread technologies that are "appropriate" to the supplier, rather than the recipient. As a result, the diffusion either is biased or simply does not occur. Fitzgerald (373) has found that U.S. agricultural technologies in Mexico benefitted only farmers already Americanized in process. While Donaldson (67) maintains that local users refuse innovations that are inappropriate to local needs, this is not generally accepted.

More frequently, however, the tendency to focus on the recipient in diffusion studies results in a prevailing concern with domestic conditions for successful or unsuccessful diffusion. Instead of viewing given technologies as inappropriate to the societies of developing countries, the societies are viewed as inappropriate to the technologies. Along these lines, Ruttan and Hayami (338) argue that "cultural endowments" play a significant role in the diffusion of technology. According to Biggs and Clay (358), the R&D unit in the developing country is most strongly affected by its work climate, and the work climate depends primarily on socio-cultural factors. Chepaitis (140) points out the cultural factors that impede the transfer of computer technologies. Gore and Lavaraj (380) suggest that the spread of new technologies is hampered when the population of the recipient country is split along some line (such as an ethnic or ideological division). Ghazanfar (963) and Goodell (378) both hold that social change is necessary for the successful adoption of innovations.

In addition to cultural barriers to diffusion, social conditions may affect the spread of new technologies through their administrative and economic settings. Bernardo (356) has concluded that the problems of gasifier diffusion in the Philippines are institutional, rather than technical. According to Quiroga (404), new institutions are required to ensure the accrual of benefit to targeted groups. Pitt and Sumodiningrat (403), in contrast to many of the critics of the Green Revolution (such as Grabowski 303), have found a reduced likelihood of the adoption of high-yield varieties by large landowners. According to Hicks and Johnson (382), the acceptance of labor-using high-yield techniques has followed population growth. Descriptions of economic conditions affecting diffusion have been offered by Antle and Crissman (269), Burstein (279), Feder and O'Mara (372), Katz (388), and Chudnovsky (362).

Even when developing country social conditions are taken as given, the emphasis of diffusion studies is still on the recipient. The problem may be defined as finding the most efficient way for diffusion mechanisms to function in a developing country setting. This sometimes is approached from an economic perspective. Lichtman (397) stresses the importance of basing diffusion efforts on an understanding of rural resource flows and local political economies.

The efficient functioning of diffusion mechanisms involves making foreign technology a part of local life. Eastman (370) has looked at how Peru managed to adopt the U.S. land-grant model (cf. Rogers 452). Koppel (393) has held that technology must be "indigenized" before it is widely diffused. "Indigenization" means that the needs of local users (most often farmers, since developing countries are largely agricultural) must be determined and met (cf. Hyman 385;

Flora 431; Piniero and Trigo 402).

Taking local conditions into account can lead to more rapid diffusion, since the diffusion follows established lines of communication. Thus, Wunsch (412) found a role for traditional leaders in expediting diffusion. Effective communication often is considered to be a key to meeting user needs, in particular two-way communication between users and diffusers (Agarwal 350; Colle 364; Gamser 375; Ali 120; Aina 353).

Since a major concern in diffusion literature is how to increase the rapidity with which technologies spread, the proper design of diffusion mechanisms is a major issue. What kinds of organizational designs are successful and what kinds are unsuccessful? Although attempts at general theories of innovation diffusion exist (such as Szyliowicz 408), the more common approach is to attempt to identify successful elements in specific situations, leading to a heavy reliance on case studies. For example, Hill (171) examined unsuccessful attempts to construct subcontracting networks to facilitate technology diffusion in the Philippines. Gamser (375) attempted to identify the characteristics of a Sudanese program that enabled it to develop initiatives in charcoal production.

One characteristic in the design of a successful diffusion program is the connection of R&D structures and formal education to informal education programs. Colle (364) and Compton (365) deal with the transfer of knowledge from scientists in R&D programs to other sectors. Biggs and Clay (358) have pointed to the existence of both formal and informal R&D systems and argued that these must work together for effective diffusion.

Similarly, Warren (462) examines interactions among scientists, extension staff, and indigenous producers. For Aina (353), the link between extension programs, a type of informal education, and formal education should be provided by librarians, who can provide information to extension officers.

These organizational and educative approaches to diffusion assume that new techniques may be spread by communication and that, therefore, organizational structure and linkages are of primary importance. Alternatively, one can hold that new techniques spread in response to economic conditions, and that creating the correct set of incentives is more important than creating the proper organizational mechanism. Pitt and Sumodiningrat (403) have found that education is not as significant in promoting the adoption of high-yield varieties as economic incentives. Thus, Piniero and Trigo (402) have examined economic inducements to adoption of innovations.

Although the study of the spread of new techniques in developing countries generally casts the recipient in the relatively passive role of accepting technology and promoting its use as widely and quickly as possible, some authors do recognize that a high state of successful diffusion can place the recipient in the active role of developing its own technology. Teitel (409) and Teubal (411), among others, have discussed the technological creativity of developing countries.

Conclusion

The tendency to regard technology transfer and technology diffusion as two separate stages in the exchange and diffusion process (cf. D'Ambrosio 726) has led to different orientations, giving rise to distinct sets of themes. When discussing the transfer of technology, the literature is generally organized, implicitly or explicitly, around the nature of the exchange relationship. If the relationship between supplier and recipient is seen as inherently unequal and disadvantageous for the recipient, this leads to a concentration on the problematic nature of the transfer. This may lead one to deny the value of technology transfer, or to stress its risks for recipients.

If it is understood as risky, but necessary, the point of concern may be strategies for the minimization of risk and for the improvement of the developing country's bargaining position. Such strategies include protectionism and South-South trade, as well as an emphasis on technology "appropriate to the environment and social order of the developing country."

Scholars less critical of supplier-recipient relations emphasize ways of maximizing benefits from transfer, rather than minimizing risks. Because the supplier can play a positive role, ways for supplier organizations to act effectively are often considered. The point of view of the supplier does not, however, appear to dominate the literature. Conflicts between the two parties may be seen as resolvable to the advantage of both. Profit seeking on the part of both parties may provide an analytical focus. By maximizing the benefits of transfer, developing countries may achieve economic development. "Appropriate technology" from this point of view tends to refer to technology that has the greatest economic viability. Definitions of appropriate technological policy, like definitions of appropriate technology, are different for different orientations. If the transfer relationship is positively valued, a policy based on trade is more likely to be advocated. Developmental difficulties are traced to domestic infrastructures, rather than to trade relations.

The social effects of transfer are most often considered by its critics, who tend to find undesirable consequences. In part, this may be because those who assume the advantages of technology tend not to seek to prove their assumptions. However, some have found beneficial social effects, such as improved living standards and higher employment.

Although diffusion may mean diffusion among nations or across national boundaries, more frequently it refers to the spread of technology among users within a nation. Thus, it occurs after the transfer and constitutes a separate stage for the purposes of the literature. Diffusion literature is dominated by the underlying question of how new technologies may achieve the greatest rapidity and density of spread. Factors that impede this rapidity and density tend to receive primary consideration. Because the focus is within the developing country, domestic conditions, such as socio-cultural, administrative, or economic barriers, receive the greatest attention. In addition to the difficulties presented by domestic conditions, the need to make new technologies a part of those conditions is recognized.

Studies of diffusion mechanisms are directed toward designing more effective means of spreading new technologies. This means that the identification of characteristics of successful and unsuccessful diffusion programs is important. The case study offers a means of comparing these characteristics.

Providing informational links among R&D organizations, formal education, and such informal educational organizations as extension agencies is often suggested as a way to improve diffusion. However, those who find economic conditions more influential than education stress the creation of economic incentives for adoption of innovations. The end result of successful diffusion is occasionally considered, but this receives less attention than solving problems in the process.

Carl L. Bankston III

Chapter 2 -- Appropriate Technology

53. Ahiakpor, James C.W. 1989. "Do Firms Choose Inappropriate Technology in LDCs?" ***Economic Development and Cultural Change*** **37: 557-71.**

Empirical study of 297 manufacturing firms in Ghana in 1970. Looks at different types of firms (private foreign, private local, private mixed, state owned, and state-foreign co-owned) to see how they rank according to capital intensity, skill-mix, and import dependence. Mixed state-foreign firms have the least "appropriate" techniques, with the highest capital-labor ratio, skill-mix, and import dependence.

54. Ahmad, Aqueil. 1989. "Evaluating Appropriate Technology for Development: Before and After." ***Evaluation Review*** **13: 310-9.**

A pragmatic and readable assessment of AT. Main issue for developing countries is in evaluating technologies in terms of cost, quality, scale, degree of sophistication, risk of failure and environmental risks. Criteria for evaluation have not been developed. The AT movement has romanticized the problem and is no longer valid. Pre-implementation assessment is needed, but conflict-of-interest problems stand in the way for developing countries.

55. Alauddin, Mohammad and Clem Tisdell. 1988. "The Use of Input-Output Analysis to Determine the Appropriateness of Technology and Industries: Evidence from Bangladesh." ***Economic Development and Cultural Change*** **36: 369-92.**

Aims to identify appropriate industries for Bangladesh, based upon availability and need of labor and capital for each industry. Comes to the qualified conclusion that agricultural industries seem to be most appropriate for Bangladesh. Suggests that Bangladesh's past urban-industrial development bias has been inappropriate.

56. Alpine, Robin and James Pickett. 1980. "More on Appropriate Technology in Sugar Manufacturing." ***World Development*** **8: 167-74.**

Consideration of David Forsyth's argument (in *World Development*, Vol. 5, March 1977) that capital-intensive sugar technologies are more profitable to Third World countries than labor-intensive technologies. Evidence from a recent study of sugar technology choices in Ghana, Kenya, Ethiopia, and Egypt. Technical discussion of the advantages of vacuum pan versus improved open pan sugar technologies. Finds that, with regard to the sugar industry, efficiency and equity may not conflict.

57. Baily, Mary Ann. 1981. "Brick Manufacturing in Colombia: A Case Study of Alternative Technologies." ***World Development*** **9(2): 201-13.**

A comparative international study of manufacturing firms that finds size and capital intensity to be greater among LDC enterprises than among those in more advanced cultures. Includes a case study of the Colombian brick industry. The rate of diffusion for superior technologies in LDCs is not merely slower than in the more developed areas. Rather, argues that the diffusion model does not fit the situation in Colombia.

58. Bhalla, A.S. (ed.). 1979. ***Towards Global Action for Appropriate Technology*****. Oxford: Pergamon Press.**

Contains papers that attempt to answer key questions about appropriate technology: (1) What are the criteria for technologies appropriate to a particular development strategy? (2) What role do existing national and subregional institutions play in the development of appropriate technologies? (3) What mechanisms exist in the UN to ensure cooperation of different agencies? (4) What mechanisms exist for networking, information dissemination, and promotion of R&D, and how effective are they? (5) Is there a case for a new international mechanism to promote appropriate technology?

59. Bhalla, A.S. 1979. "Technologies Appropriate for a Basic Needs Strategy." Pp. 23-62 in ***Towards Global Action for Appropriate Technology*****, edited by A.S. Bhalla. Oxford: Pergamon Press.**

Assumes that developing countries should be concerned with rapid fulfillment of the basic needs of poorer populations, obtained through increasing national and collective self-reliance within the framework of a New International Economic Order. Links appropriate technology concepts and basic needs concepts by giving a technological content to the basic needs approach, then examines the demands such an approach might make on decentralization of production structures. Also considers the political and administrative requirements for technological self-reliance in developing countries.

60. Bhalla, A.S. and D.D. James. 1986. "Technological Blending: Frontier Technology in Traditional Economic Sectors." *Journal of Economic Issues* 20: 453-62.

Defines technological blending as the constructive integration of newly emerging technologies with traditional economic activity. Calls for the conscious instigation of carefully preplanned projects designed, implemented and carried out within the conceptual and analytical framework of technological blending. Three types of projects are especially recommended: (1) those that include a central means of providing and dispersing technology to traditional sectors, (2) those that augment existing resources, and (3) exploratory projects that seek open-ended technological blending.

61. Bhalla, A.S. and D.D. James. 1991. "Integrating New Technologies with Traditional Economic Activities in Developing Countries: An Evaluative Look at Technology Blending." *Journal of Developing Areas* 25: 477-96.

Outlines several case studies of technology blending, the integration of cutting-edge technologies with traditional production methods. Part of the attention is dedicated to the social consequences of these innovation methods. The article is optimistic about the possibility of introducing new technologies into traditional economies in a "symbiotic" fashion.

62. Bowonder, B. 1979. "Appropriate Technology for Developing Countries: Some Issues." *Technological Forecasting and Social Change* 15: 55-68.

Obtaining self-organizing systems without huge inputs of energy and information is difficult. Seven major limitations to the analysis of appropriate technologies for developing countries: absence of formalized criteria, exogenous planning, existence of multiple needs, complexities of the future, value dependence of technology, information gaps, and poor policy-making capabilities. An integrated view of AT has not been developed; it is "subjective, undefined, qualitative in nature, and depends mainly upon ... convenience" (p. 60). Twenty potential criteria for selecting ATs are given.

63. Bruch, Mathias. 1983. "Technological Heterogeneity, Scale Efficiency, and Plant Size: Micro Estimates for the West Malaysian Manufacturing Industry." *Developing Economies* 21: 267-77.

Finds that a switch to small-scale production would increase employment and cause a serious decrease in productivity. Intermediate technologies can mitigate that decrease somewhat.

64. Chatel, B.H. 1979. "Technology Assessment and Developing Countries." *Technological Forecasting and Social Change* 13: 203-211.

Activities of the UN with respect to the assessment of technologies in the late 1970s are reviewed. Main criteria for assessment of technology in developing countries are employment, scarce capital, and energy; design of appropriate technologies should use criteria of taste, culture, needs, purchasing power, and raw materials.

65. Davies, D.M. 1985. "Appropriate Information Technology." *International Library Review* 17: 247-58.

Tries to define "appropriate information" and looks at some of the criteria for determining appropriate info technology. Gives succinct discussion of the world science information system (UNISIST), developed by UNESCO, and examples of libraries and information centers in developing countries.

66. Dickson, David. 1974. *The Politics of Alternative Technology*. New York: Universe Books.

Originally published as *Alternative Technology and the Politics of Technical Change*. Discusses the social function of technology and its socially accepted legitimations. Argues that the problems of technology arise from its nature and its social uses, that technology can never be considered independent of social and political factors, and that alternative technology can be developed which would avoid these problems. Genuine alternative technology can be developed only within another societal framework. Chapters on problems with contemporary technology in both industrialized and underdeveloped countries, how technological innovation is political, and particular types of alternative technology, including one 25-page chapter on intermediate technology for the Third World.

67. Donaldson, William. 1980. "Enterprise and Innovation in an Indigenous Fishery: The Case of the Sultanate of Oman." ***Development and Change*** **11: 479-95.**

Locals are not opposed to accepting innovations, and when they do reject "innovations" imposed from above it is often because the innovation wasn't appropriate.

68. Dunn, P.D. 1978. ***Appropriate Technology: Technology With a Human Face.*** **London: Macmillan.**

A positive portrayal of AT with an emphasis on a number of specific genres of technology--food and agriculture, energy, water, health, medicine, construction, transportation. Discusses problems of education, training, and R&D, and offers strategies for "getting started."

69. Eckaus, Richard S. 1987. "Appropriate Technology: The Movement Has Only a Few Clothes On." ***Issues in Science and Technology*** **3: 62-71.**

First part of a compelling debate with Frances Stewart on the value of the concept of appropriate technology. Eckaus proposes a variety of reasons that the idea of appropriate technology is misleading and may not be in the interests of developing countries at all.

70. Ellis, Gene. 1981. "Development Planning and Appropriate Technology: A Dilemma and a Proposal." ***World Development*** **9: 251-62.**

Examines the system of project paper preparation and project planning by the U.S. Agency for International Development and other donor agencies. Discusses three approaches for remedying the weaknesses of the system: a "holistic" approach to the planning problem, an "operant conditioning" approach to implementation, and a "process" approach to planning and implementation. On the basis of these three approaches, the author suggests an approach grounded in the search for an approximately appropriate technology.

71. Fleissner, P. (ed.). 1983. *Systems Approach to International Technology Transfer*. Oxford: Pergamon Press.

Based on a 1983 Symposium of the International Federation of Automatic Control. Takes an appropriate technology perspective. General introductory papers on systems approaches to AT transfer, papers on technology transfer in selected fields, case studies, flexible manufacturing systems, energy technologies, information and communication, and economic, educational and social aspects. One of the principal conclusions is that developing countries should put more emphasis on informatics.

72. Floor, Willem. 1979. "Activities of the UN System on Appropriate Technology." Pp. 138-163 in *Towards Global Action for Appropriate Technology*, edited by A.S. Bhalla. Oxford: Pergamon Press.

Reviews activities of the UN on appropriate technology. Finds that only a small part of UN science and technology activities are concerned with AT, that activities of different organizations overlap considerably due to lack of systematic planning, and that coordination is mostly at the country-project level.

73. Forsyth, David J.C. 1977. "Appropriate Technology in Sugar Manufacturing." *World Development* 5: 189-202.

A detailed economic and engineering survey of potential technologies for sugar refining. Finds that capital-intensive technologies are clearly the most economically efficient except at the smallest levels of production. Comment and reply in volume 7 (# 8-9) and volume 8 (pp. 165-6).

74. Forsyth, David J.C., Norman S. McBain and Robert F. Solomon. 1980. "Technical Rigidity and Appropriate Technology in Less Developed Countries." *World Development* 9: 371-98.

Study finds that for five countries, labor-intensive technologies were used when they were available, but have little overall effect in reducing unemployment, causing the authors to question the likely potential of ATs in lowering unemployment.

75. Griffith, W.H. 1990. "CARICOM Countries and Appropriate Technology." *World Development* 18: 845-58.

Argues that capital-intensive technologies are more appropriate for the development of export-oriented technology in the Caribbean Community (CARICOM) countries than labor-intensive technologies. The author reviews the arguments in favor of an export-oriented strategy for CARICOM countries: (1) the population of CARICOM is only about 6 million, providing insufficient domestic demand, (2) the distribution of income is skewed toward those with a low propensity to consume, further limiting demand, (3) traditional export agriculture, sugar and bananas, is unable to provide sufficient jobs, and (4) the external indebtedness of CARICOM countries is heavy.

Regional governments have accepted these arguments, but export-oriented and import-substitution industrialization have not substantially reduced underemployment and unemployment. Critics attribute this to the failure to use appropriate (labor-intensive) agriculture. The author examines the leading export-oriented sectors--agriculture, bauxite, petroleum, and manufacturing. He maintains that in each of these areas, the region must adopt the most recent, capital-intensive technology, in order to compete with the developed countries.

76. Hamelink, Cees. 1988. *The Technology Gamble: A Study of Technology Choice*. Norwood, NJ: Ablex Pub. Corp.

Takes gambling as a model for technology choice and argues that many developing countries have been overcome by "the compulsive neurosis of obsessive gambling." From this point of view, the author argues that lack of responsibility in technological choice should be criminalized. Revision of technology choice should be based on flexibility of planning, scope for intervention in technology projects (including the possibility for dismantling projects that turn out to have undesirable effects), and continuing assessment procedures in order to learn from errors.

77. Henry, Paul Marc, Amulya K.N. Reddy and Frances Stewart. 1979. "A Blueprint for Action." Pp. 207-220 in *Towards Global Action for Appropriate Technology*, edited by A.S. Bhalla. Oxford: Pergamon.

Presents a plan for appropriate technology action based on consultations held by the authors in a cross-section of countries and at the headquarters of various intergovernmental organizations.

Recommends that a special Founders' Conference of the UN should be organized to determine the functions, location, and funding of a new mechanism to promote appropriate technology. This new mechanism should be a nongovernmental institution, outside the body of the UN, but closely associated with it through a sponsorship arrangement.

78. Hetman, Francois. 1977. "Social Assessment of Technology and Some of Its International Aspects." *Technological Forecasting and Social Change* 11: 303-14.

Outlines a new method for social assessment of technology.

79. Howes, Michael. 1979. "Appropriate Technology: A Critical Evaluation of the Concept and the Movement." *Development and Change* 10: 115-24.

General review of Schumacher's concepts and the AT movement, criticizing it for neglecting the larger social, political, and economic context.

80. Hungwe, Kedmon. 1989. "Culturally Appropriate Media and Technology: A Perspective from Zimbabwe." *TechTrends* 34: 22-3.

Discusses options in Zimbabwe for development of culturally appropriate media and technology, given the country's limited resources and poverty. Concludes that the issue of appropriate media is dependent on the quality of the personnel and that therefore the emphasis should be on training of teachers and media personnel.

81. Huq, M.M. and H. Aragaw. 1977. "Technical Choice in Developing Countries: The Case of Leather Manufacturing." *World Development* 5: 777-89.

A comparative study of the evolution of leather manufacturing techniques in Argentina, Brazil, India and Pakistan, examining the range of technological choice available to LDCs. Concludes that (1) developing countries desiring to establish a leather manufacturing industry have a wide range of technological choice and (2) technologies using more labor and less fixed capital appear to be better choices than more capital-intensive methods.

82. Jequier, Nicolas. 1979. "Appropriate Technology: Some Criteria." Pp. 1-22 in *Towards Global Action for Appropriate*

***Technology*, edited by A.S. Bhalla. Oxford: Pergamon Press.**

Spells out criteria that should be taken into account in guiding the selection of appropriate technologies. Defines appropriate technology as small-scale technology that has its origins in traditional methods of production. Presents criteria such as cost, risk, modernity, individual or collective nature, and single or multipurpose character of technology. Sees non-economic problems (sociological and institutional constraints) as more important in hindering the adoption of appropriate technologies than purely economic considerations.

83. Jequier, Nicolas and Gerard Blanc. 1983. *The World of Appropriate Technology: A Quantitative Analysis*. Paris: OECD.

Seeks to provide comprehensive data on appropriate technology activities throughout the world and to present a general picture of the AT movement at the end of the 1970s. Examines the quantitative growth of appropriate technology organizations, attempts to answer basic questions about the fields of activity of AT organizations, and analyzes the funding and staffing of AT organizations. Examines obstacles to diffusion of innovations (mainly financial). Analyzes and maps the communications networks in AT. Tries to develop a typology of AT institutions. The main conclusion is that AT is entering its second generation and facing new problems.

84. Kaniki, Martin. 1980. "Economical Technology against Technical Efficiency in the Oil Palm Industry of West Africa." *Development and Change* 11: 273-84.

Locals refused technology offered them by technocrats, but subsequent studies found that the locals made the correct choice.

85. McRobie, George. 1979. "Intermediate Technology: Small Is Successful." *Third World Quarterly* 1: 71-86.

Schumacher's intellectual descendent and author of *Small Is Possible*. Offers a number of examples of success.

86. Ndonko, W.A. and S.O. Anyang. 1981. "Concept of Appropriate Technology: An Appraisal from the Third World." *Monthly Review* 32: 35-43.

Rejects the suggestion that a technology can be "appropriate" or otherwise. Points to the actual pattern of socio-economical development in a community as the key factor--a technology is or is not "appropriate" based upon whether it fits into the ongoing schema of development. Therefore the AT concept is a tautology: an "appropriate" technology is any technology that is appropriate.

87. Onn, Fong Chan. 1980. "Appropriate Technology: An Empirical Study of Bicycle Manufacturing in Malaysia." *Developing Economies* 18: 96-115.

Maintains that developing countries must conduct extensive research to determine the types of technology most appropriate to their environment. One way of doing this is to examine techniques for manufacturing a particular product to identify a technique most appropriate with regard to an accepted set of criteria. To this end, bicycle manufacturing in Malaysia is investigated.

Assumes that criteria of appropriateness depend on goals for development and that the principal development goal of Malaysia is eradication of poverty with a reduction of inequality of income distribution. Detailed consideration, with statistics on the Malaysian bicycle industry, of Raleigh and six other bicycle manufacturing firms. Concludes that an improved version of the technology of the Far East firm may be the most appropriate, while the technology of Raleigh, which is a foreign-controlled firm, may not be appropriate. At the end, briefly discusses the pattern of income poverty in Malaysia and how utilization of AT for bicycle manufacturing will be consistent with the government's goal of eradicating poverty.

88. Ovitt, G. 1989. "Appropriate Technology: Development and Social Change." *Monthly Review* 40: 22-32.

A "committed" engineer at Drexel discusses sample AT projects as an alternative to MNC "tech-transfer," the political ideas behind it, a bit of history, and many current examples.

89. Parthasarathi, A. 1990. "Science and Technology in India's Search for a Sustainable and Equitable Future." *World Development* 18: 1693-1702.

Challenges the idea that developing countries can or should mimic development patterns similar to those followed by currently industrialized countries. An alternative development pattern sensitive to lifestyles, settlement patterns and productive systems can be based upon alternative technologies in foodgrain production, energy, science-based rural technologies and biotechnology.

90. Radnor, Michael and Stephen Kaufman. 1988. "Facing the Future: The Need for International Technology Intelligence and Sourcing." Pp. 305-312 in *Science, Technology, and Development*, edited by Atul Wad. Boulder: Westview Press.

Argues that LDCs must look to political means to redress the power imbalance in acquisition of technology. They also must rely upon their endogenous capacity-building technology and on horizontal transfers within the developing world, for the following reasons: (1) technology needed by developing countries is owned by large MNCs who will only make it available on unfavorable terms of cost and dependency, (2) much of this technology is not appropriate to developing country needs, and (3) finding, evaluating, selecting, and acquiring the right technology is expensive and not feasible for most LDCs. Developing countries need a formal department dedicated to providing technology intelligence and finding appropriate technology.

91. Reddy, Amulya Kumar N. 1975. "Alternative Technology: A Viewpoint from India." *Social Studies of Science* 5: 331-42.

Suggests that India's stark division between haves and have-nots has a technological basis, resulting from importation of "Western technology." Western technology is capital-intensive and labor saving in a nation that needs just the opposite; it also produces a wealthy sector at the expense of the rest of the population. Adds that domestic technological innovation is not the answer. Rather, alternate technology based upon the nation's human resources and village-style communities should be developed or sought. Warns that an impending oil crisis might force a move to decentralized production.

92. Reddy, Amulya K.N. 1979. "National and Regional Technology Groups and Institutions: An Assessment." Pp. 63-137 in *Towards Global Action for Appropriate Technology*, edited by A.S. Bhalla. Oxford: Pergamon.

Argues that developing countries already have S&T institutions that give them national capabilities. But these institutions are not applied to the development and dissemination of appropriate technology, and they usually do not have direct contact with users. As a consequence, the needs of small-scale users have been neglected.

93. Riedijk, W. 1982. "Appropriate Technology for Developing Countries: Toward a General Theory of Appropriate Technology." Pp. 3-20 in *Appropriate Technology for Developing Countries*, edited by W. Riedijk.

Challenges assumption that technology is "value-free." Mass production means that anonymous markets must be created, exports must be emphasized, and products must touch all levels of society. The "technostructure" provides for overproduction and elite dominance.

94. Riskin, Carl. 1978. "Intermediate Technology in China's Rural Industries." *World Development* 6: 1297-1311.

Reviews two decades of experimentation with intermediate technology in Chinese rural industries. Gives a brief history of technical choice policy in the PRC. Attempts to identify some of the functions of intermediate technology, with reference to employment creation, imperfect factor markets, suboptimal savings and farm labor demand, the policy of technological dualism ("walking on two legs": trying to develop both a rural industrial sector and a modern capital-intensive type of industry), and self-reliance. The author concludes that China's rural intermediate industrial technologies are appropriate to their conditions and tasks.

95. Robinson, Austin (ed.). 1979. *Appropriate Technologies for Third World Development.* London: Macmillan.

A collection of essays addressing AT experiences on a comparative national basis, including China, Japan, the Philippines, Taiwan, Turkey, and Iran. Also includes essays discussing broader topics, such as the role of multilateral financial agencies, aid donors and MNCs, as well as general theoretical problems.

96. Salmen, Lawrence F. 1987. *Listen to the People: Participant-Observer Evaluation of Development Projects*. Oxford: Oxford University Press.

A World Bank publication outlining experiments in Mexico using participant observers to evaluate development projects. Those affected by development know which projects are working and which aren't long before administrators do. The P-O method is a means to facilitate communication.

97. Schumacher, E.F. 1973. *Small Is Beautiful: Economics As If People Mattered*. New York: Harper and Row.

Classic work which first used the term "Intermediate Technology."

98. Schumacher, E. F. 1981. "Buddhist Economics." Pp. 326-34 in *Technology and Man's Future*, edited by Albert Teich.

Reprint of influential article proposing the "humanizing" of technology. Suggests that the important task for technological and economic development is to find a balance between "materialist heedlessness" and "traditionalist immobility." Schumacher took Ne Win's Burma as a model of Buddhist economics. However, the sensibility of the article, the feeling that spiritual well-being should accompany economic progress, continues to be the source of various ecological perspectives, such as "deep ecology."

99. Sharif, M. Nawaz and V. Sundararajan. 1984. "Assessment of Technological Appropriateness: The Case of Indonesian Rural Development." *Technological Forecasting and Social Change* 25: 225-38.

Tests the authors' method of evaluating a technology's "appropriateness" using two Indonesian rural development experiments, one emphasizing the difficulty of identifying a "set of ATs" and the other stressing problems in selecting "the most appropriate one" for any village.

100. Smil, Vaclav. 1976. "Intermediate Energy Technology in China." *World Development* 4: 929-38.

A general description of the successful use of AT in China, including biogas and hydroelectricity.

101. Stewart, Frances. 1977. *Technology and Underdevelopment*. Boulder: Westview Press.

Classic investigation of the relationship between income inequality and technological dependence. Includes critique of conventional theory of technological choice and discussions of employment problems in poor countries, the characteristics of advanced country technology and their distorting effects on development, the nature of AT, nature and consequences of technological dependence, the role of capital-goods industries, and international trade. Followed by 3 chapters on micro questions: choice of technique, maize grinding and cement-block manufacture in Kenya. Technological dependence comes from reliance on foreign technology and concentrated sources. Undesirable consequences of dependence: (1) cost, (2) loss of control over decisions, (3) unsuitable characteristics of technology received, (4) lack of effective, indigenous scientific and innovative capacity. Advocates trade among Third World countries and development of a local capital-goods industry.

102. Stewart, Frances. 1979. "International Mechanisms for Appropriate Technology." Pp. 164-206 in *Towards Global Action for Appropriate Technology*, edited by A.S. Bhalla. Oxford: Pergamon.

Reviews international mechanisms, both existing and proposed, for dealing with appropriate technology. Concludes that existing mechanisms do little to further AT. Maintains that international action is needed to promote appropriate R&D on appropriate technology.

103. Stewart, Frances (ed.). 1987. *Macro-Policies for Appropriate Technology in Developing Countries*. Boulder, CO: Westview Press.

A collection of essays that serve as case studies of how technology was chosen in several areas: India, Bangladesh, Thailand, the Philippines, and Kenya (all rural technologies) as well as Kenya, Tanzania, and Latin America (for urban and industrial technologies). The editor's introduction argues that political and economic macropolicies strongly influence the choice of technologies in LDCs, and that these influential policy instruments can be altered to promote ATs.

104. Stewart, Frances. 1987. "The Case for Appropriate Technology: A Reply to R.S. Eckaus." *Issues in Science and Technology* 3: 101-9.

Surveys the arguments in favor of appropriate technology in a response to R.S. Eckaus' critique in the same issue.

105. Teitel, Simon. 1977. "On the Concept of Appropriate Technology for Less Developed Countries." *Technological*

Forecasting and Social Change **11: 349-70.**

By one of the first economists to recognize indigenous technological effort in LDCs. Presents a survey of development experts that suggests two hypotheses: (1) LDCs do not merely adopt technology passively; (2) technology is often adapted to the local environment. Offers alternative reasons why technology is "inappropriate" for a community, mostly commercial rather than cultural.

106. Thoburn, John T. 1977. "Commodity Prices and Appropriate Technology--Some Lessons from Tin Mining." ***Journal of Development Studies*** **14: 35-52.**

Evaluates the "appropriateness" of two techniques for tin mining introduced into Malaysia, using cost-benefit criteria. The rankings, determined both by private and social profitabilities, are found to be sensitive to the product price in addition to the discount rate, leading the author to offer his results as an argument in favor of schemes to stabilize primary commodity prices.

107. Trak, Ayse and Michael MacKenzie. 1980. "Appropriate Technology Assessment: A Note on Policy Considerations." ***Technological Forecasting and Social Change*** **17: 329-38.**

Proposes an alternative approach to AT assessment. To the traditional capital- and consumer-goods sectors, the authors add a "technological innovation" sector. Importation of technology by a country should be conceived as a contribution to the third sector, not just the capital-goods sector.

108. VanDam, A. 1979. "Third World Collaboration: Determining the Appropriate of Appropriate Technology." ***Futurist*** **13: 65-7.**

Report on growing technical cooperation among Third World countries by a delegate to a UN conference on the issue in December 1978. Gives examples of contemporary technical cooperation, seeks a definition of appropriate technology, looks at technical education, and gives one project as an example. A very brief overview of all these topics.

109. Westphal, Larry. 1978. "Research on Appropriate Technology." ***Industry and Development*** **2: 28-46.**

Seeks to clarify AT issues and identify areas where further research is required to improve design of technology-choice policy in the manufacturing sector. Finds that research on appropriate technology in manufacturing has tended to concentrate on the technical feasibility of capital-labor substitution. Because no simple rule exists to achieve the appropriate choice in this substitution, technological choices are complex and must be made according to specific circumstances. Therefore, research into policies and institutional mechanisms should have high priority.

110. White, L.J. 1978. "The Evidence on Appropriate Factor Proportions for Manufacturing in Less Developed Countries." *Economic Development and Cultural Change* 27: 27-59.

Attempts a systematic analysis of evidence on the existence of efficient alternatives to Western technology for developing countries. Summarizes arguments concerning the absence of choice in selecting efficient factor proportions and offers a survey of the available evidence. Explores the possibility of developing new products and processes to widen the range of technological choice and increase productivity. Concludes that greater labor intensity in manufacturing for developing countries is feasible and would be efficient. Evidence suggests that incentives, such as appropriate factor prices, are important. Policy implications: establishment of proper factor prices is very important; efforts to provide special facilities or subsidies to small firms probably are unwise; the policy toward used machinery should be relatively neutral; tougher bargaining by LDCs together with appropriate factor price and procompetition policies would limit abuses by MNCs; research institutes and information services are probably good ways of encouraging more appropriate R&D by and for LDCs.

111. Willoughby, K.W. 1990. *Technology Choice: A Critique of the Appropriate Technology Movement.* Boulder, CO: Westview.

Calls for policies with Appropriate Technology movement principles stressing the importance of technology choice. But rejects limiting "technology choice" to acceptance of a certain set of technologies, describing it instead as a method for decision making.

112. Winston, Gordon C. 1979. "The Appeal of Inappropriate Technologies: Self-Inflicted Wages, Ethnic Pride and Corruption." *World Development* 7: 835-46.

Surveys a number of Nigerian manufacturing firms for clues explaining why LDC companies often choose "inappropriate" technologies. Finds that technology decisions are not always made according to neoclassical production models, but instead are made according to individual firm managers' concepts of AT and technology cost. In many cases firm managers considered non-objective reasons when making inappropriate technology choices, including defense of ethnic identity or personal enrichment.

113. Wong, Christine P.W. 1986. "Intermediate Technology for Development: Small-Scale Chemical Fertilizer Plants in China." *World Development* 14: 1329-46.

Embodying "appropriate technology," these plants use domestic equipment and older water-gas methods to produce low-analysis fertilizers for local use. Built during the 1960s and 1970s, they originally were uneconomical. Relative prices changed, making them viable but only transferable to countries with small markets and low-cost local fuels and feedstocks.

114. Yapa, Lakshman S. 1982. "Innovation Bias, Appropriate Technology and Basic Goods." *Journal of Asian and African Studies* 17: 32-44.

The goals of Third World economic development can be met only by producing basic goods first. Reasons why AT is essential to the success of basic goods development strategy: (1) Allocation of resources for basic needs will require major political effort; AT has an important role to play in organization and maintenance of that work. (2) Research and development of technology are basic mechanisms used in the reproduction of social relations of production. (3) Mass production of basic goods becomes possible only through use of AT, because these technologies require far less capital than conventional ones. (4) Universal provision of basic goods must be ecologically sustainable in the long run, which is only possible through use of environmentally appropriate tech. Basic goods include a biogenic diet, clothing, shelter, health care, and functional literacy. Absence of goods at one level of society is causatively linked to the resource consumption of non-basic goods at another level of society. Drastic restructuring of political and economic power is necessary.

Chapter 3 -- Technology Transfer

115. Agmon, Tamir and Mary Ann von Glinow (eds.). 1991. *Technology Transfer in International Business.* New York: Oxford University Press.

Contains 14 articles, all concerned in various ways with the overlap between international business and technology transfer. Technology transfer is seen as an integral part of international business. Three main sets of essays: The Environment of Technology Transfer, Some Elements of Technology Transfer, and The Practice of International Technology Transfer in the Pacific Rim. Articles dealing with specific countries are all concerned with the U.S., Japan, and China, except for chapter 12 (Pros and Cons of International Technology Transfer: A Developing Country's View), which is concerned with Korea.

116. Akpakpan, Edet B. 1986. "Acquisition of Foreign Technology: A Case Study of Modern Brewing in Nigeria." *Development and Change* 17: 659-76.

Turn-key approach to technology acquisition resulted in independence in the operation of beer production but dependence on outside engineering necessary for industrialization of brewing.

117. Al-Ali, Salahaldeen. 1991. "Technological Dependence in Developing Countries: A Case Study of Kuwait." *Technology in Society* 13: 267-78.

Sources, types, elements, methods, and selection of technology transfer from the U.K. to Kuwait. Turn-key agreements and supply of machinery were found to be the main methods for transferring foreign technology to Kuwait. These led to technological dependence, since almost 90% of the supplier firms were Western owned and controlled and only two firms were jointly owned by Kuwait and Britain. The local work force has not been trained much, leading to a shortage of engineers. Kuwait must turn its attention to the development of an indigenous technological infrastructure.

118. Alam, Ghayur and John Langrish. 1981. "Non-Multinational Firms and Transfer of Technology to Less Developed Countries." ***World Development*** **9: 383-8.**

Examines non-MNCs to determine their usefulness as sources of technology transfer in relation to MNCs. Finds they are adequate but overlooked because of MNC prestige.

119. Alange, Sverker. 1987. ***Acquisition of Capabilities through International Technology Transfer: The Case of Small Scale Industrialization in Tanzania.*** **Goteborg, Sweden: Chalmers University of Technology.**

Looks at the acquisition of static and dynamic capabilities in small-scale Tanzanian industries, with small-producing firms in Sweden as technology suppliers under the Sister Industry Programme. The data were collected from 1978 to 1986, and come from interviews with 22 Tanzanian firms and 18 Swedish firms, as well as from interviews with consultants and authorities. Acquired capabilities were studied in depth in 1983, in order to assess the technology transfer program. Small-scale firms were successful at transferring production capabilities, but less successful at transferring administrative and dynamic capabilities. Over the long term, however, the Tanzanian firms did acquire dynamic capabilities. The study makes recommendations for improving the sister-industry type of transfer.

120. Ali, S. Nazim. 1989. "Science and Technology Information Transfer in Developing Countries: Some Problems and Suggestions." ***Journal of Information Science*** **15: 81-94.**

The author, from the University of Bahrain, examines the link between technology transfer, R&D, and information access. Effective transfer to promote development can only take place when R&D activities exist within the country, and this can occur only with effective information transfer. Summary of problems and techniques in scientific information transfer: lack of hard currency, libraries and information centers, absence of national bibliographies, and available media (hard-copy, on-line access to databases, microform, optical disks).

121. Arnon, I. 1989. ***Agricultural Research and Technology Transfer.*** **London: Elsevier.**

Very broad overview of the role of technology transfer in

agricultural research. Covers a wide variety of countries and historical periods. Focus is agricultural research systems, human resources, and physical resources needed for efficient technology transfer in agricultural research. Includes advice on how to manage and administer research projects, how to evaluate research proposals, and potentially negative effects of research, such as environmental problems and unemployment. References are thorough and will be helpful to anyone interested in agricultural research management.

122. Asheghian, P. 1985. "Technology Transfer by Foreign Firms to Iran." *Middle Eastern Studies* 21: 72-9.

Investigates problems of technology transfer by foreign firms to Iran, concentrating on technology adaptation, labor training, and research and development.

123. Baark, Eric. 1987. "Commercialized Technology Transfer in China 1981-86: The Impact of Science and Technology Policy Reforms." *China Quarterly* 111: 390-406.

Reforms of 1980s were meant to solve a persistent problem in China since early 1950s: technology transfer from research to production units. Most research is performed in independent state-operated units, but the emergence of the contract system and technology markets in the early 1980s increased the supply of technology without increasing the demand for it. Contracts are important instruments of decentralized control. Reports have surfaced in the Chinese press that improved research productivity and rates of diffusion have been achieved. Establishment of technology fairs and a patent law in 1985 gave rights to ownership of technology. Recently, stronger bureaucratic control has been exerted, causing a slowdown of the technology market. Ill feeling results from redistribution of income through sale of technology. Low demand for technology is created by a risk-aversive enterprise leadership. There is a lack of facilities for engineering development at many research units. Average level of R&D expenditures at Chinese enterprises is less than 1% of output value.

124. Baranson, Jack. 1970. "Technology Transfer through the International Firm." *American Economic Review* 60: 435-440.

Examines factors influencing logistics and conflicts of interest in licensing versus investment decisions. Four sets of interrelated factors are: (1) complexity of product and production techniques being transferred, (2) transfer environment in donor and recipient countries, (3) the absorptive capacities of the recipient firm, (4) the transfer capability and the profit-maximizing strategy of the donor firm. All influence preferences of international firms for direct investment or licensing.

125. Baranson, Jack. 1978. *Technology and the Multinationals.* Lexington, MA: Lexington Books.

A series of case studies intended to illustrate a new development in the international setting for technology sharing--that is, the growing willingness of U.S. corporations to sell industry technology to "noncontrolled foreign enterprizes" (technology sharing). Studies include the aircraft, automotive, computer, consumer electronics, and chemical engineering industries. Draws policy implications from the technology-sharing trends found.

126. Baranson, Jack and Robin Roark. 1985. "Trends in North-South Transfer of High Technology." Pp. 24-42 in *International Technology Transfer: Concepts, Measures, and Comparisons*, edited by N. Rosenberg and C. Frischtak. New York: Praeger.

Looks at the definition of high technology and problems of measurement. Reviews the shift toward high technology among NICs, including upward movement in the international division of labor, the shift from import substitution to export promotion, and national policies for technologically intensive transfer--the patterns of response to demand and problems in tech transfer.

127. Batra, Raveendra N. and Sajal Lahiri. 1987. "Imported Technologies, Urban Unemployment and the North-South Dialogue." *Journal of Development Economics* 25: 21-32.

High royalty rates on imported technologies reduce the ability of imported agricultural technologies to increase rural employment. On the other hand, reduced royalties on urban industrial technologies would not help decrease unemployment, and might even increase it.

128. Beaumont, C., J. Dingle and A. Reithinger. 1981. "Technology Transfer and Applications." *R&D Management* 11: 149-56.

The conditions necessary for success in international tech transfer are seldom in tune with trends in political aspirations. Recipient organizations have nationalistic objectives (profitable future business) while transferrer wishes to make immediate profit. Most tech transfer does not have to do with the goal of development. Good brief review of international political framework (including UN codes), the commercial framework, and the operational framework of ITT. Developing countries are most in favor of a legalistic approach to regulation of TT that actually hurts them (offers a case study of a turn-key contract supporting this view). All technologies are appropriate if nested in a viable system (especially one that is not in decline relative to the international system).

129. Behrman, Jack N. and W.A. Fischer. 1980. *Overseas R&D Activities of Transnational Corporations*. Cambridge, MA: Oelgeschlager, Gunn, and Hain.

Not specifically about developing countries, but a good study of multinational R&D activities. Based on interviews with 56 companies in the U.S., Japan, and Europe in 1978, the focus is on the R&D activities of foreign affiliates of TNCs. Discussion of the types of R&D performed, how they were selected, how R&D is managed, the kinds of collaborative international activities involved, the diffusion of R&D-related capabilities, and the impacts of governmental policies on R&D location and performance. The meat of the book is in three lengthy appendices, somewhat older case studies of product lines by Johnson & Johnson (health care), Unilever (international R&D coordination), and DuPont (fabrics and finishes).

130. Behrman, Jack N. and Harvey W. Wallender. 1976. *Transfer of Manufacturing Technology Within Multinational Enterprises*. Cambridge, MA: Ballinger.

Compilation of case studies of technology transfer in manufacturing, based on interviews with management and affiliates. MNCs studied were Ford Motor Co., International Telephone and Telegraph, Pfizer Corporation, and Motorola. Host countries were South Africa, Taiwan, Mexico, Nigeria, Brazil, and Korea. Introductory chapter includes a technology-transfer matrix of types and mechanisms of transfer but authors seek description, not generalization. Little attempt to draw together the broad lessons learned from the case studies.

131. Bell, Martin and Don Scott-Kemmis. 1988. "Technology Import Policy: Have the Problems Changed?" Pp. 30-70 in *Technology Absorption in Indian Industry*, edited by Ashok V. Desai. New Delhi: Wiley Eastern Limited.

Based on the results of a British/Indian study of technology transfer. Information was collected in detailed interviews with 42 British firms that had transferred technology to Indian firms. Suggests that the problems of the 1960s and 1970s, which Indian technology import regulations were designed to answer--such as control of international markets by monopolistic MNCs--have changed as a result of diversification of supplier firms and other developments. By the late 1970s and early 1980s, the Indian policy regime was having a negative effect on modernity of technologies.

132. Bommer, Michael R.W., Ralph E. Janaro, and Deborah C. Luper. 1991. "A Manufacturing Strategy Model for International Technology Transfer." *Technological Forecasting and Social Change* 39: 377-90.

Discusses the need for transfer technology that can be adapted to resource conditions in recipient countries to assure appropriateness and achieve competitiveness. Extends a manufacturing strategy model developed by T.J. Hill to encompass the factors important to the strategic decisions of international technology transfer. Offers a systems view of the ITT model.

133. Boye, T. and A. Fenichel. 1988. "Skill Transfer and African Development: A Conceptual Research Note." *Journal of Modern African Studies* 26: 685-90.

Holds that a major weakness in the debate between development theories and basic needs approaches is the general acceptance of the view that the transfer of technology can be reduced to the transfer of machines, products, and processes. In the light of the African experience, this presumption is unwarranted. The article attempts to identify major factors affecting skill transfer and to suggest possibilities for further research.

134. Bruton, Henry J. 1977. "A Note on the Transfer of

Technology." *Economic Development and Cultural Change* 25: 234-44.

Develops a mathematical model for relating employment growth to productivity growth, substitutability among inputs, and other factors. Suggests how the model may be applied. Emphasizes the adaptation of technology to the resource endowment of the host country over time, such that foremen and managers drive technological development more than scientists.

135. Burch, David. 1987. *Overseas Aid and the Transfer of Technology*. Brookfield, VT: Avebury.

Examines the role of overseas aid in Sri Lankan tractorization. Part 1 is a history of British overseas aid and technology transfer, an excellent introduction even for those not interested in Sri Lankan tractorization. Part 2 focuses on tractorization, including a study of the world tractor industry. Tractorization was encouraged by the donor country as means of expanding exports. The mechanization of ploughing benefitted rich farmers at the expense of poor farmers. The study is critical of agricultural mechanization in general, and of the British-supported agricultural mechanization in Sri Lanka in particular.

136. Capriles, Roberto Salas. 1977. "Technology Transfer and the Industrialists in Latin America." *Impact of Science on Society* 27: 307-20.

How can Latin America develop its own technology? Emphasizes the need for entrepreneurship and research. Transfer of technology is a process in which a country is free to choose autonomously from different alternatives.

137. Chanaron J.J. and J. Perrin. 1987. "The Transfer of Research, Development and Design to Developing Countries: Analysis and Proposals." *Futures* 19: 503-12.

Examines the problems associated with incorporation of research, development and design facilities into schemes to transfer manufacturing technology to developing countries. Contrasts two models that can be used for evaluating the modalities, conditions, and impact of such transfers.

138. Chatterje, M. 1990. *Technology Transfer in the Developing*

***Countries*. NY: St. Martin's Press.**

A collection of essays divided into four parts: (1) Conceptual Issues in Innovation and Technological Diffusion, (2) Strategic Dimensions of Technology Transfer to Developing Countries, (3) The Use of Specific Technologies in Economic Development, and (4) Some National Case Studies of Technology Transfer. Case studies include Southeast Asia, China, Singapore, and Nigeria. Industries include microcomputers, nuclear energy, ocean technology, and robotics.

139. Chen, Edward. 1981. "The Role of MNCs in the Production and Transfer of Technology in Host Countries." *Development and Change* 12: 579-99.

MNCs have helped Hong Kong because of willingness to train workers and import technology.

140. Chepaitis, Elia. 1990. "Cultural Constraints in the Transfer of Computer Technologies to Third World Countries." Pp. 61-71 in *International Science and Technology: Philosophy, Theory and Policy*, edited by Mekki Mtewa. New York: St. Martin's Press.

A general examination of cultural factors that can impede transfer of computer technologies. Suggests that the first step toward successful transfer of computer technology is identification of cultural constraints. The second step is to determine which internal changes are desirable or unavoidable, and which changes should be prevented. The final step is to ask whether the transfer itself is possible and desirable.

141. Clark, Norman. 1975. "The Multinational Corporation: The Transfer of Technology and Dependence." *Development and Change* 6: 5-21.

Explores the proposition that access to and use of differentiated technical know-how is an important factor in the ability of the MNC to maintain control over internal economic production in developing countries. Suggests that product cycle theory is an unsatisfactory descriptor of modern international techno-economic relations, but that advanced technology does play a role in the behavior of MNCs. The MNC is itself a product of historic technical change, because MNCs are operating in industrial sectors in which R&D and innovation play a key production strategy role, and because they are able to establish monopolistic positions in host economies as a result of combined market and technological power. The author concludes that, because possession of and access to modern technology has been important in enabling MNCs to dominate developing countries, any attempt to break free of the dependent relationship must be predicated on development of an independent, indigenous technical base closely linked to economic production.

142. Clarry, John William. 1987. "The Regulation of Technology Transfer by Multinational Corporations." ***Dissertation Abstracts International, A: The Humanities and Social Sciences*** **48: 1330-A.**

Ph.D. dissertation (State University of New York at Stony Brook, 1986). Examines the changing terms and organizational forms of technology transfer by MNCs within the pharmaceutical industry in Latin America. Predictions of dependency were confirmed, but these were moderated by multinational strategies and by bargaining by elites of host governments.

143. Cole, W.E. and J.W. Mogab. 1987. "The Transfer of Soft Technologies to Less-Developed Countries: Some Implications for the Technology/Ceremony Dichotomy." ***Journal of Economic Issues*** **21: 309-20.**

Study of two Mexican projects (based on the Tennessee Valley Authority) that sought to transfer and adapt foreign institutional arrangements to Mexican settings. Veblen-Ayres paradigm is reviewed and revised. Relative independence of the TVA is contrasted with two Mexican commissions; the more independent commission was more successful. Concludes that in practice, technological and ceremonial behavior are not absolutely opposed: ceremonial behavior can be important in progress.

144. Compton, J. Lin (ed.). 1989. ***The Transformation of International Agricultural Research and Development.*** **Boulder & London: Lynne Rienner Publishers.**

Commissioned by the Experiment Station Committee on Organization and Policy (ESCOP). Looks at the influence of US State Agricultural Experiment Stations in educating foreign students, developing new concepts and research results, and transferring knowledge and technology to developing countries. Divided into three parts: historical perspectives on context and change in agricultural development, selected problems, and future challenges.

145. Cooper, Charles. 1980. "International Transactions in Innovative Machinery." ***Journal of Development Studies*** **16: 332-51.**

Looks at the economic implications of two different types of transactions, "machine maker" and "machine user" transfers, to bring new technology into Third World industries. Suggests that type of transfer is less important in terms of social costs and benefits than level of government mechanisms, such as price controls, to retain control. Concludes that Third World countries will not necessarily be better off if they obtain new technologies directly from machine makers rather than from machine users.

146. Cortes, Mariluz and Peter Bocock. 1984. ***North-South Technology Transfer: A Case Study of Petrochemicals in Latin America.*** **Baltimore: Johns Hopkins University Press.**

A descriptive study of the transfer of petrochemical technology to a group of industrializing countries in Latin America, with a nice awareness of theoretical questions regarding transfer of technology from developed to developing countries. The introduction gives good summaries of two theories of tech transfer (the technology gap theory and the product life-cycle theory), while the body of the book is a detailed examination of petrochemical suppliers and recipients. Concludes that (1) the main determinants of type of contractual arrangements, in order of importance, are type of product, recipients' country, and suppliers' country; and (2) product and process maturity factors are not significant in determining the type of technology agreement used.

147. Dahlman, Carl J. and Larry E. Westphal. 1981. "The Meaning of Technological Mastery in Relation to Transfer of Technology." ***Annals of the American Academy of Political and Social Science*** **458: 12-26.**

Acquisition of technological mastery is critical to self-sustaining development. Technology transfers are substitutes for local mastery rather than sources. To what extent can effective use be made of available knowledge without an indigenous effort to master it? Turn-key contracts are the most frequent mode of transferring technology for activities that are entirely new to an economy--they may fail to supply the recipients with full understanding. Conclusion--the dependence of an economy's fund of technical expertise on mastery means initial decisions about choices of technology and degrees of local involvement are critical determinants of the direction in which the economy's technological mastery will develop.

148. Davidson, William. 1980. *Experience Effects in International Investment and Technology Transfer*. Ann Arbor, MI: University of Michigan Press.

Study of foreign direct investment and licensing activity by U.S.-based companies. Analysis focuses on basic investment decisions: location, timing and use of independent licensees, joint ventures or wholly owned subsidiaries as recipients of foreign manufacturing for a product. Results of tests indicate that experience is a powerful variable in explaining the international spread of manufacturing for the sample of products. Examines the distinction between general and specific experience. Also confronts a rival hypothesis for explaining the acceleration in foreign manufacturing that suggests this activity is due to increasing competition for global markets rather than the experience of companies. The finding that significant innovations spread as quickly as other new products suggests that competition is not the principal factor in foreign manufacturing trends. Looks at these findings in light of corporate policy, international investment and public policy.

149. Davidson, William. 1983. "Structure and Performance in International Technology Transfer." *Journal of Management Studies* 20: 453-65.

Formal macro-organizational structure is important in determining technology transfer because it affects the ability of firms to learn from previous experience. Structures centralizing learning benefits (i.e., an international division) are more efficient, while firms organized by global product divisions exhibit slower and lower levels of transfer activity than firms in other structures (duplication of transfer mechanisms). Sample of 57 U.S.-based MNCs generated 954 new products, which was the unit of analysis. Measures of technology transfer and structural forms are discussed.

150. de Aragao, Paulo Ortiz Rocha and Reeve Vanneman. 1990. "Technology Transfers and Managerial-Professional Employment:

Brazilian Manufacturing, 1960-1975." ***Latin American Research Review*** **25: 87-101.**

Important attempt to resolve the debate on the long-term effect of technology transfer, using two regression analyses of data from Brazil. The first is a hierarchical regression of change in the proportion of university-trained technicians between 1960 and 1975 in Brazil. The second is a panel regression analysis of the effects of technological dependence, growth of employment, and structural variables on change in the proportion of university-trained technicians during this period. In moderate amounts, technology imports lead to increases in technically trained personnel, but beyond a certain point other processes take hold that override the positive impact of new technologies. There is thus a curvilinear relationship between new imported technologies and growth in trained personnel.

151. Derakshani, Shidan. 1984. "Factors Affecting Success in International Transfers of Technology: A Synthesis and a Test of a New Contingency Model." ***Developing Economies*** **22: 27-46.**

Technology transfer is defined as the acquisition, development, and utilization of technological knowledge in a country other than that in which this knowledge originated. Decision makers in developing countries are faced with a bewildering set of decisions--after the choice of technology, they must go about structuring their relationship with the technology source to best achieve their objectives in the transfer process.

The success of technology transfer can be judged operationally, according to returns on investment (ROI), the level and growth of sales. Or transfer can be judged according to its second-order consequences: (a) international competitiveness as indicated by exports, (b) indigenization and reliance on domestic personnel inputs, (c) ability of the enterprise to innovate. Also must be seen from points of view of recipient and supplier. Model describes chance of success as greater when the supplier is more involved. Such motivation is greater in a relationship of greater supplier control, greater interaction, more initial involvement, and more stability.

152. Desai, Ashok. 1984. "India's Technological Capability: An Analysis of its Achievements and Limits." ***Research Policy*** **13: 303-310.**

Argues that India's success in importing technologies, reliant in part of the ability to build plants, varies according to national demand in the industry's market. A good or service with less attraction on the national market is less likely to survive without heavy investment, promoting inefficient and unreliable facilities.

153. deVries, Peter. 1988. "Technology Transfer: Introducing a Computer to Teach Number Skills to Adults in Soweto." ***Convergence: An International Journal of Adult Education*** **21: 5-15.**

Describes the introduction of a system of computer-assisted instruction at one adult learning center in Soweto in South Africa to teach basic numeracy skills to adults.

154. Dietz James L. and Dilmus D. James (ed.). 1990. ***Progress Toward Development in Latin America: From Prebisch to Technological Autonomy*****. Boulder: Lynne Rienner Publishers.**

Articles on structuralist economics, dependency, import substitution, and the role of science and technology in development.

155. Edoho, F.M. 1990. "Technology-Transfer and Third-World Development: The Petrochemical Complex in Nigeria." ***Bulletin of Science, Technology, and Society*** **10: 201-11.**

Case study of Nigeria's petrochemical complex, intended to illustrate the dilemmas of Third World development strategies that rely heavily on technology transfer. The first part contrasts two theoretical frameworks of development--capitalist theory and "critical social theory" (generally dependency theorists or world systems theorists). The second part is a case study of technology transfer to Nigeria's petrochemical complex. Tables in this section include a ranked listing of oil MNCs active in the Nigerian economy, Nigeria's crude oil production from 1958 to 1982, and government joint ventures (with percent of government participation) with MNCs in the oil industry. The third part is an assessment of the contribution of oil to Nigeria's economy, focusing on the transfer of oil technology by the MNCs. The author maintains that Nigeria's lack of integration in the world economy and dependence on oil technology transfer from the West limits its ability to pursue autonomous development strategies. The last parts examine the feasibility of developing indigenous oil technology, seen as critical to Nigeria's development, and conclude that technology development is not purely technical or economic, but includes political, social, and cultural dimensions.

156. Enos, John. 1984. "Government Intervention in the Transfer of Technology: The Case of South Korea." ***IDS Bulletin*** **15(2): 26-31.**

Assesses government role in the importation of petrochemical technology, and finds a positive effect.

157. Enos, J.L. and W. H. Park. 1988. ***The Adoption and Diffusion of Imported Technology: The Case of Korea.*** **London: Croom Helm.**

Examines attempts to adopt modern technology in South Korea for a period of about 20 years (study conducted in the late 1970s and updated in mid 1980). Focuses on the petrochemical, synthetic fiber, machinery, and iron and steel industries, with detailed analysis by a team of an economist and an engineer in each industry. Examines the techniques available, the reasons for choosing a technology, the means of transfer, the efficiency of absorption, changes made, and consequences for development. The organizations studied are independent Korean firms. Contains chapters on methodology, economic background, and a comparison of the Korean experience with Japan's. Concludes that the role of the Korean government has been crucial.

158. Fangyi, Huang. 1987. "China's Introduction of Foreign Technology and External Trade: Analysis and Options." ***Asian Survey*** **27: 577-94.**

Author from the Institute of World Economy and Politics at Chinese Academy of Social Sciences. Surveys China's introduction of foreign technology and external trade after 1979. Main problems: overcontrol, duplication, inefficiency, "stagnated diffusion."

159. Fritz, Jack H. 1988. "Microcomputers, Dams and Handpumps: Three Case Studies of Technology Transfer to Developing Countries." Pp. 211-226 in ***Science, Technology and Development,*** **edited by Atul Wad. Boulder: Westview Press.**

Compares the diffusion of three types of technology in developing countries. Discusses several trends affecting the technology transfer process: the trend toward independence from food aid among developing countries, the debt crisis, the non-competitiveness of U.S.-manufactured products, the emphasis on private-sector initiatives, and the emphasis on security assistance.

In looking at microcomputer diffusion, author argues that (1) microcomputers will "inundate" developing countries, and that governments should encourage efficient use of the technology, (2) since the microcomputer has no attendant dangerous effects, its use should be encouraged, (3) by using microcomputers, developing country professionals will become familiar with a variety of modern technologies, (4) the technology offers an opportunity for developing countries to get involved in defining their own uses and sources, since there are many opportunities for assembly and software development, (5) the microcomputer is not an active agent of development, but a passive one in which humans must act.

Handpumps are described as designed for rural use directly by intended beneficiaries. Therefore, handpumps must fit into their cultural framework, and the role of women must be central. The author is critical of large dams, and suggests that they are usually based on technical and economic aspects, and built by political elites without consideration for the effects on the public. Some strategies for reducing negative health impacts of dams are suggested.

160. Georgantzas, Nicholas C. and Christian N. Madu. 1990. "Cognitive Processes in Technology Management and Social Change." *Technological Forecasting and Social Change* 38: 81-95.

Development of a cognitive mapping model (Influence Diagramming) to assist teams of LDC planners and MNC managers in making decisions and dealing with complexity, ambiguity, and uncertainty. In the process of transfer, opposing world views between MNC and LDC managers become the source of "productive conflict"--process of conflict resolution leads to "bargaining windows" of innovation and change. Conflict is necessary--events at Bhopal show danger of achieving goal compatibility too soon. "Negentropy" acheived when technology transfer is driven by LDC-MNC compatible goals. Appropriate technology is important where the preservation of institutional arrangements takes precedence.

161. Girvan, Norman. 1978. "White Magic: The Caribbean and Modern Technology." *Review of Black Political Economy* 8: 153-66.

A lucid and critical discussion of the Caribbean experience with technology transfer that compares faith in the developmental capability of technology with superstitious beliefs in magic. Current "technology transfer" is criticized as a misnomer, since it is expensive, doesn't deliver know-how, often comes with subsidiary licensing deals, and bears hidden costs. Successful modes of "technology transfer" must be more selective, drawing from diverse countries and emphasizing human resource needs.

162. Greenfield, Howard. 1988. "Information Technology Transfer and the Developing Nations of the Asian Pacific Rim." ***Microcomputers for Information Management*** **5: 129-36.**

General and somewhat elementary consideration of prospective benefits from applying new computer-based information technologies in this geographic area. Discusses the deterrents to implementation, and whether these technologies can help the Asian Third World modernize. Gives very brief examinations of the information technology situation in North Vietnam, China, Singapore, and Malaysia.

163. Gritzner, Charles F. 1981. "Technology Transfer to the Third World: Boon or Bane?" ***Journal of Geography*** **80: 192-93.**

Questions the value of technology transfer for people and traditional cultures in the Third World. Concentrates on environmental and social damage. Of little use for those who already have a general familiarity with social and environmental problems of technology transfer.

164. Grynspan, D. 1982. "Technology Transfer Patterns and Industrialization in LDCs: A Study of Licensing in Costa Rica." ***International Organization*** **36: 795-806.**

Studies licensing as a technology transfer mechanism in Costa Rica in order to examine the role of foreign investment in import-substitution industrialization and its contribution to the development of Central American economies. Based on answers given to questionnaires administered to general or production managers of 30 firms producing under license during the second half of 1979. Concludes that the data are not sufficient to evaluate the net impact of MNC activity in Costa Rica, but licensing as a mode of technology transfer by MNCs was found to be less than adequate in promoting development.

165. Hamelink, Cees. 1985. "High Tech Transfer: Selling the Canoe without the Paddle." *Development* 1: 28-37.

Transfer of technology is basically the transfer of knowledge/information. Information technology is the command and control system for all other technology. Information technology is the fourth largest industry in the world--the chief architect of infrastructures through which values of the transnational business system are spread globally. Today the main beneficiaries of transnational data flows are located in North America, Western Europe and Japan. "As is the case with technology in general, real transfer of information technology rarely takes place. What is usually transferred are end products and related services." Primary beneficiaries are foreign or national elites, and the interests of the North and South are at odds. Suggests a "Southern strategy"--intra-south transfers with technological choices made in line with defined development objectives.

166. Hanson, John R. 1989. "Education, Economic Development, and Technology Transfer: A Colonial Test." *The Journal of Economic History* 44: 939-957.

Challenges the idea that the importation of modern technology and prospects for Third World economic development are a function of the local population's formal schooling, by showing that repayment for education in the manufacturing sector is even lower than repayment in mining.

167. Hawkins, John N. 1988. "The Transformation of Education for Rural Development in China." *Comparative Education Review* 32: 266-81.

Contrasts the "transfer" model (linear transfer of knowledge from urban areas to periphery) to the "transformation" model (back and forth linkages with multiple centers of diffusion). Agricultural reforms made since 1978, such as changes in rural educational policy, have led to transfers from urban to rural sectors and locally based transformations. A main problem is the perpetual tendency for the government to recentralize reforms.

168. Headrick, Daniel R. 1988. *The Tentacles of Progress:*

***Technology Transfer in the Age of Imperialism, 1850-1940.* New York: Oxford University Press.**

Looks at the role of various types of technology in relations between imperial Britain and its colonies, with some attention also given to France and its colonies. Areas considered include shipping, railways, telecommunications, sanitation, irrigation, botany and plantations, mining and metallurgy, and technical education. Concludes that technological change was a mixed blessing for both colonized and colonizers: for the colonizers because technological advancement was the basis of their power, and a source of profit, but brought social change and helped undermine their rule; for the colonized because technologies were associated with the foreign rulers and because they were at a competitive disadvantage, but also provided the only avenue to advancement. Excellent historical background to contemporary North-South technological relations, with a wealth of empirical detail in every chapter. Bibliographical Essay at the end will be valuable to anyone interested in the history of technology transfer.

169. Helleiner, G.K. 1988. "Transnational Enterprises in the Manufacturing Sector of the Less Developing Countries." Pp. 203-223 in *Technology Transfer By Multinationals*, edited by H.W. Singer, Neelambar Hatti, and Rameshwar Tandon. New Delhi: Ashish Publishing House.

The prices paid for foreign technology are not fixed, but depend on bargaining. Looks at bargaining strategies for LDCs.

170. Heller, Peter B. 1989. "International Student Mobility: Some Consequences of a Form of Technology Transfer." *International Studies Notes* 14: 36-40.

The article notes that about a third of a million foreign students, mostly from the Third World countries, are studying at American universities. It then questions the consequences, in particular observing how these students affect transfer of skills between the U.S. and their own countries. Begins with a brief examination of the growth of student exchange over the past 40 years. Looks at the types of skills transferred to the students--language skills and cultural information, as well as technical, scientific, and managerial skills. Foreign students have reduced the parochialism of American business and the students have benefitted from the American analytical and critical approach to problems.

The foreign students may be "sold" on American values, or their exposure to American life may reinforce stereotypes and leave them ambivalent toward the U.S. Observes that student exchange has been a kind of "reverse dependency," providing a critical mass of students to American universities at a time of shrinking enrollments. Gives examples of restraints on technology transfer through student exchange, such as restricting access of Warsaw Pact students to supercomputers, reluctance of some governments to send students abroad, mismatch between American training and home country technology, ostracism of returning students for seeming too "American." Concludes that everyone benefits from student exchange.

171. Hill, Stephen. 1986. "Eighteen Cases of Technology Transfer to Asia/Pacific Region Countries." ***Science and Public Policy*** **13: 162-9.**

Summarizes a provoking set of technology transfer cases, searching them for lessons. Tech transfers arising from aid, commercial transactions, and government research institutes. Some technologies are needed and some are imposed by the politics of aid or the personal wishes of leaders; others are incompatible with existing local technological infrastructure. Different levels of local technological decision-making capability noted along with difficulty in establishing influential social ties for science and technology planners in government. Concludes with a checklist of factors to take into account when planning tech transfer.

172. Hope, K.R. 1983. "Basic Needs and Technology Transfer Issues in the New International Economic Order." ***The American Journal of Economics and Sociology*** **42: 393-403.**

New perceptions of development are examined as they relate to basic needs and the transfer of technology. This "basic needs" approach seeks to reduce and eventually to eliminate dependence on developed country enterprises, thus allowing developing countries to control their natural resources. Seeks to accelerate self-reliance and to introduce some measure of global management of resources.

173. Hruza, Edward and Harry G. Miller. 1983. "Technology Adaptation in Developing Countries: An Intervention Model."

***Journal of Studies in Technical Careers* 5: 195-204.**

Outlines a decision-making model with different "levels" of intervention in developing countries (health and automotive examples used).

174. International Labor Organization. 1988. *Safety, Health, and Working Conditions in the Transfer of Technology to Developing Countries*. Geneva: ILO.

A code to follow in setting up technology transfer to LDCs. Intended to protect Third World workers from hazardous conditions.

175. Johnston, Ann and Albert Sasson (eds.). 1986. *New Technologies and Development*. Paris: UNESCO.

Based on a 1984 UNESCO colloquium covering general trends and specific applications of new technologies. Separate treatments on information technology, biotechnology in farming and food systems, health care technologies and delivery systems, energy technologies, and educational change. Papers include discussion of the array of new technologies involved in each, together with the scientific basis, and issues and priorities raised by each for developing countries. Includes an introduction on technology in society, the diversity of conditions, and obstacles to development. Limited bibliography.

176. Kaimowitz, David (ed.). 1990. *Making the Link: Agricultural Research and Technology Transfer in Developing Countries*. Boulder: Westview Press.

Seven essays on the relationship between agricultural research and technology transfer, including a focus on political economy, state policy, the private sector, and conceptual frameworks.

177. Kaplinsky, Raphael. 1976. "Accumulation and the Transfer of Technology: Issues of Conflict and Mechanisms for the Exercise of Control." *World Development* 4: 197-224.

Suggests that technology transfer always involves a conflict between the interests of the supplier and the receiver, and that the stronger usually wins. Also discusses the alliance between control and choosing the type of capital accumulation.

178. Kaplinsky, Raphael. 1988. "Export-Oriented Growth: A Large

International Firm in a Small Developing Country." Pp. 466-87 in *Technology Transfer By Multinationals*, edited by H.W. Singer, Neelambar Hatti, and Rameshwar Tandon. New Delhi: Ashish Publishing House.

Empirical study of a large, export-oriented international firm operating a pineapple factory in Kenya. Distribution of gains favors the company rather than Kenya.

179-180. Karake, Zeinab A. 1990. *Technology and Developing Economies: The Impact of Eastern European versus Western Technology Transfer*. New York: Praeger.

Through analysis of the Middle East experience, argues that technology transfer occurs within a broader foreign policy context. U.S. technology transfer has a higher emphasis on security and military concerns, while Eastern European nations are found to place greater emphasis on "manpower training" than Western nations. But Eastern European technologies are found to have poor performance in the Middle East, probably because of their capital-intensive nature and generally poor quality. Appendix includes data on shares of suppliers in the imports of the countries in the sample. Contains good bibliography of works on development and technology transfer, although many are somewhat dated.

181. Katrak, Homi. 1989. "Imported Technologies and R&D in a Newly Industrialising Country: The Experience of Indian Enterprises." *Journal of Development Economics* 31: 123-40.

Empirical study of Indian firms (data from 1966-71, top 300 firms; and 1980-84, 51 firms) shows that imports increased the chances a firm would begin R&D. Firms which spend more on imports also spend more on R&D. Imports of technology help promote in-house R&D, but the effect is limited. Larger firms have proportionately lower R&D spending.

182. Katrak, Homi. 1990. "Imports of Technology and the Technological Effort of Indian Enterprises." *World Development* 18: 371-82.

Examines whether the import of technology by developing-country enterprises encourages or discourages technological effort, in the light of the Indian experience. Multiple regression analysis indicates that where imports are intended to complement technological efforts, these imports encourage the further development of local technology. If developing country enterprises have obtained an exclusive right of sale in the home country, however, the import of technology is found to have a negative effect on technological development.

183. Kilby, Peter. 1979. "Evaluating Technical Assistance." ***World Development*** **7: 309-23.**

Reviews "institutional factors" that have prevented technical assistance project evaluation, such as (1) the difficulty of measuring a given project's actual contribution of employment or value, (2) the extreme cost in capital and physical labor of attempting to measure the contribution, and (3) the occasionally significant time lapse between project end and project evaluation. Proposes a methodology by which case studies can be evaluated and offers advice from experience concerning choices of activity, project design and project management.

184. Kim, Linsu. 1980. "Stages of Development of Industrial Technology in a Developing Country: A Model." ***Research Policy*** **9: 254-77.**

Presents a three-stage model of industrial technology development, to study how characteristics of international tech transfer, the production process structure, and local firms' innovation strategies evolve in response to the changing competitive climate. The model is based on primary data collected from 31 consumer- and industrial-electronic product manufacturers as well as secondary data on the electronics industry in Korea. In the first stage, implementation of a production operation was the primary concern of firms, with product design and production technologies generally imported. In the second stage, the focus was on assimilation of foreign technologies. In the third stage firms concentrated on the improvement of foreign technologies. Discusses the applicability of this model to other industrial sectors in Korea and offers a set of propositions on industries in developing countries.

185. Kim, Linsu. 1991. "Pros and Cons of International Technology Transfer: A Developing Country's View." Pp. 223-239 in ***Technology Transfer in International Business*****, edited by Tamir Agmon and Mary Ann von Glinow. New York and Oxford: Oxford**

University Press.

Assesses pros and cons of international technology transfer, presents a conceptual framework to identify and assess different forms and channels of international technology transfer, and examines how an LDC acquires technology from DCs using Korea as an example. In the conceptual framework, modes of technology transfer are seen as market-mediated or nonmarket-mediated. Market-mediated transfer may involve active suppliers, in the form of direct foreign investment, foreign licensing, turn-key plants, technical consultancies, and made-to-order machinery. Foreign suppliers may play a passive role as in standard machinery purchase. Nonmarket-mediated transfer, which involves an active foreign partner, may take the form of technical assistance by foreign buyers or technical assistance by foreign vendors. Foreign partners play a passive role in nonmarket transfer in imitation, trade journals, and technical information service. Gives a brief discussion of Korean policies for technology transfer, beginning in the 1960s. These are examined using the market-nonmarket, active-passive frameworks. Gives implications for developed and developing countries.

186. Kollard, F. 1990. "National Cultures and Technology Transfer: The Influence of Mexican Life-Style on Technology Transfer." ***International Journal of Intercultural Relations*** **14: 319-36.**

Reports results of a questionnaire given to 100 TNC executives concerning the influence of Mexican culture and daily life upon technology transfer. Develops three models of how Mexican culture was perceived: (1) the hedonist, (2) the refuser, (3) the uncertain. Argues that a TNC executive's adoption of these three Mexican stereotypes influences his general approach to technology transfer as well as whether technical renewals and employee training are planned.

187. Kumar, Nagesh. 1987. "Technology Imports and Local Research and Development in Indian Manufacturing." ***Developing Economies*** **25: 220-33.**

Investigates the role of imported technological devices on local R&D. Finds that local firms importing technology through licensing tend to complement the imports with their own research, while those importing through packaged forms (FDI) tend not to do their own

R&D.

188. Lall, Sanjaya. 1980. "Brandt on 'Transnational Corporations' Investment and the Sharing of Technology'" *Third World Quarterly* 2: 701-5.

Outlines part of the Brandt Report, dealing with technology and TNCs. The author shares the optimism of this report, but maintains that it ignores trends such as the increasing inclination toward export-led, TNC-dependent development. Suggests that Third World Development is taking place within the TNC framework and therefore sees little need for a "programme of survival."

189. Lall, Sanjaya (ed.). 1984. "Special Issue." *World Development* 12: 471-660.

This special edition surveys technology exports from Hong Kong, Taiwan, Korea, India, Egypt, Brazil, Mexico, Argentina and Latin America.

190. Lall, Sanjaya. 1987. *Learning to Industrialize: The Acquisition of Technological Capability in India*. London: Macmillan.

One section from the large World Bank study of the acquisition of technological capability (by Dahlman and Westphal) in India, South Korea, Brazil, and Mexico. Focuses on cement, iron and steel, and textile industries (16 manufacturing and five consulting firms). Outlines an analytical framework, the Indian policy setting, and the sample firms. Analyzes development in the three industries, along with government policy and the success of these developments. Finds that India has become a technology-exporting country. Uncertain whether the export of technology has occurred at the expense of export of products are mixed. The conclusion attempts to draw generalizations for Third World technology acquisition from the Indian case.

191. Long, Frank. 1979. "The Role of Social Scientific Inquiry in Technology Transfer." *American Journal of Economics and Sociology* 38: 261-74.

Maintains that economics offers too restrictive a disciplinary framework for an understanding of technology. This is illustrated with some of the interdisciplinary dimensions of technology in developing countries. Concludes that a development framework, incorporating interdisciplinarity, is a prerequisite for formulating meaningful technology policies.

192. Long, Frank. 1983. "The Management of Technology Transfer to Public Enterprises in the Caribbean." ***Technology in Society*** **5.**

Attempts to show how poor management of technology transfer by public enterprises hurts both the performance of these enterprises and development itself in the Caribbean. Offers a different approach to technology transfer management in its development of indigenous technology, including a discussion of inappropriate technology choice and technological dependency.

193. Long, W.J. 1991. "Economic Incentives and International Cooperation: Technology Transfer to the People's Republic of China, 1978-86." ***Journal of Peace Research*** **28: 175-190.**

Author is a political scientist and attorney specializing in international trade law practice. Looks at U.S. technology transfer to China to determine how this promoted bilateral cooperation and contributed to international peace and stability. A good brief background to U.S.-China technology relations. This includes the embargo policy between 1949-78, the China 'Tilt' from 1978-81, the implementation of relaxed U.S. policy on export to China from 1982-86, the political setting (particularly with regard to the USSR) that led to increased cooperative Sino-U.S. interaction from 1978-84.

Summarizes the limits to cooperation, such as China's repressive domestic politics, changing relations of the two powers with the USSR, and China's exports of Silkworm missiles to Iran. The China case suggests that an economic incentive's impact on stability depends upon collateral consequences to allies and adversaries, and not just on the bilateral relationship. Concludes that economic inducements may be a qualified success in promoting cooperation, but this depends on how these inducements are introduced and executed.

194. Looney, R.E. 1988. "The Impact of Technology Transfer on the Structure of the Saudi Arabian Labor Force." ***Journal of Economic Issues*** **22: 485-92.**

Finds that development of the Saudi labor force has been uneven and premature and that the abundance of capital in the kingdom has only emphasized other shortages, such as the shortage of indigenous labor and skilled workers.

195. Madeuf, Bernadette. 1984. "International Technology Transfers and International Technology Payments: Definitions, Measurement and Firms' Behavior." ***Research Policy*** **13: 125-40.**

Methodological issues related to measurement of tech transfer by technological balance of payments. This measure includes other-than-technology components and does not record other tech transfers at all. Three main measurement problems: transfer price, economic value of technology, and national significance of the flows.

196. Madu, Christian N. and Rudy Jacob. 1989. "Strategic Planning in Technology Transfer: A Dialectic Approach." ***Technological Forecasting and Social Change*** **35: 327-38.**

Offers an approach to strategic planning intended to enhance the decision-making process of both MNCs and LDCs when transferring technology. Rather than rejecting transfer in situations when MNC and LDC goals and needs conflict, the authors' framework only requires "reasonably compatible" goals, since the dynamic of conflict can synthesize mutually beneficial results.

197. Mansfield, Edwin. 1975. "International Technology Transfer: Forms, Resource Requirements, and Policies." ***American Economic Review*** **65: 372-6.**

Describes various forms of technology transfer, discusses the problems and costs involved in the transfer process, and comments on some aspects of U.S.-USSR technology transfer. Outlines an operational measure of the resource costs of tech transfer: average cost of transfer in machinery and electrical equipment is 36%, while for chemicals and petroleum refining it is 10%. Explanation of variation in costs sought in the age of technology, experience with previous plants, and the number of prior adopters. Notes that in the previous few years there had been a marked increase in technology transfer between the U.S. and the USSR and that the benefits and costs to the U.S. of increases in technology transfer are hazy.

198. Mansfield, Edwin, Anthony Romeo, Mark Schwartz, David Teece, Samuel Wagner, and Peter Brach. 1983. "New Findings in Technology Transfer, Productivity, and Development." ***Research***

Management **March-April: 11-20.**

A study of 37 chemical, semiconductor, and pharmaceutical innovations. Finds that technology is being transferred across national boundaries more rapidly than in the past. In part, this is due to the growing influence of multinational firms and makes the idea of a product life cycle less useful than it has been in the past. Discusses the outflow of U.S. technology, the costs of ITT, and the question of slowdown in the U.S. rate of innovation.

199. Marjit, S. 1990. "On a Noncooperative Theory of Technology Transfer." ***Economics Letters*** **33: 293-98.**

A mathematical proof of a non-cooperative theory. Shows that a firm with advanced technology might be able to sell the knowledge to a competing firm that previously had less advanced technology even when there is not collusion between the two firms. Demonstrates that this is most likely to occur between firms that are fairly close in their initial technologies.

200. Marton, Katherin. 1986. "Technology Transfer to Developing Countries Via Multinationals." ***World Economy*** **9: 409-26.**

Good general review of recent developments in the tech-transfer policies and experiences of developing countries. Pattern of transfer in the 1960s and 1970s shifted in the late 1970s, owing to liberalization of regulations in some countries. Types of technology are beginning to differentiate in the extent to which regulations apply to them.

201. Meissner, Frank. 1988. ***Technology Transfer in the Developing World: The Case of the Chile Foundation.*** **New York: Praeger.**

Examines the Fundacion Chile (FCh), which was founded by decree in 1976 in order to facilitate the transfer of new technologies to Chile. Describes the origins, structure, staff, and operations of the FCh, profiles of FCh projects, the transfer of marketing technologies through FCh, financing of the organization, project monitoring and performance evaluation, achievements, and FCh's relevance as a precedent for other national technology transfer organizations. This last chapter will be of interest for students of tech transfer in developing countries. Contains a very thorough bibliography of articles published about the Chile foundation abroad and in Chile, and also materials on technology transfer. Sources in this bibliography are in both Spanish and English and date from the late 1970s to the mid 1980s.

202. Merrill-Sands, Deborah and Simon Chafer. 1989. *The Technology Triangle: Linking Farmers, Technology Transfer Agents, and Agricultural Researchers.* United Kingdom: The Roman Press.

Summarizes presentations and discussions of an international workshop that reviewed the findings of two ongoing studies, conducted by the International Service for National Agricultural Research (ISNAR) in connection with national agricultural research systems. Questions how to strengthen links between farmers and technology transfer agencies. The two studies focused on five key areas: (1) agricultural development and research policies, (2) resource situation and organizational structure of the institution, (3) technical issues, (4) inventory of available technologies, (5) diversity of agroecological conditions and production systems.

The first study concerned the organization and management of on-farm, client-oriented research. Its aims were to identify common problems, to diagnose institutional factors leading to these problems, and to provide guidelines for research managers on how to develop institutional conditions which will promote effective research on farms and with farmers. The study involved case studies of Bangladesh, Ecuador, Guatemala, Indonesia, Nepal, Panama, Senegal, Zambia, and Zimbabwe. The second study involved research-technology transfer linkages. Its purpose was to analyze research-technology links in a variety of institutional contexts and to arrive at suggestions on how these links could be improved. This involved case studies of Colombia, Costa Rica, Cote d'Ivoire, Dominican Republic, Nigeria, Philippines, and Tanzania. Concise summaries of case studies are given in boxes throughout the text. Appendices list conference participants, conference agenda, members of the ISNAR study groups, publications from the ISNAR studies, and relevant acronyms.

203. Michalet, Charles Albert. 1979. "The International Transfer of Technology and the Multinational Enterprise." *Development and Change* 2: 157-74.

Since the 1960s, transfer methods have undergone important changes. Patents and licenses have declined in favor of dissemination based on MNEs, which puts suppliers and buyers in direct touch.

Subsidiaries are subject to two spheres of decision: (1) MNE is a production unit of scientific and tech knowledge, (2) MNE operates within the national scientific and technical systems. MNEs represent integrated spheres of decision making and activity. Tendency toward centralization of MNE research in home country. When R&D activity is organized at subsidiary level--a highly specialized, localized laboratory dependent on MNE. Semi-autonomous affiliates and external dissemination are due to the need of MNEs to carry out a defensive strategy that maintains an increased share of the market. MNEs try to prevent leaks of technological elements.

204. Mohan, D. 1990. "Tantrums for Technology Transfer." *Economic and Political Weekly* 25: 249-50.

Polemical article maintaining that efforts toward technological self-reliance in post-colonial societies like India and China during the 1950s and 1960s have been reversed in recent years. This is attributed to "an unholy alliance of convenience ... between the interests of capitalist nations and third world yuppies." The National Front government of India is seen as serving the desires of the upper classes for foreign luxuries by promoting technology transfer.

205. Monkiewicz, Jan. 1989. *International Technology Flows and the Technology Gap: The Experience of Eastern European Socialist Countries in International Perspective.* Boulder: Westview Press.

Argues that, for a nation to "catch up" successfully through technology transfer, three conditions must be met: (1) a national environment conducive to innovation; (2) an internal system favorable to transfer; (3) the correct intensity of technology inflow. In the Eastern European case all three of these elements are found to be missing.

206. Moore, Omar K., Francine E. Jefferson, and Marcia Gilbert-Crosse. 1991. "Heuristic Guidelines for Analyzing Technology Transfer." *Knowledge: Creation, Diffusion, Utilization* 12: 298-310.

The world is evolving into one intricately connected socio-economic system. Being on the "cutting edge" is not enough; technologies actually must be put to use. Technology transfer is the process whereby scientific discoveries and engineering achievements are transformed into useful goods and services. Gives case studies in U.S.

207. Morgan, Robert P. 1983. "Sharing Science and Technology." *Bulletin of the Atomic Scientists* 39(5): 23-7.

Questions the ability of the North's science and technology to contribute to the South's development, with a bill of specific proposals as well as more general ones.

208. Mucchielli, Jean-Louis. 1987. "Multinational Enterprises, International Investments, and Transfers of Technology: The Elements of an Integrated Approach." Pp. 11-33 in *Multinationals, Governments and International Technology Transfer*, edited by A.E. Safarian and Gilles Y. Bertin. New York: St. Martin's Press.

Examines various forms of international exchange using the ideas of comparative advantage of the country and competitive advantage of the firm. Determinants of multinationalization of production lie in the relation between the two.

209. Mytelka, Lynn Krieger. 1978. "Licensing and Technology Dependence in the Andean Group." *World Development* 6: 447-59.

Examines 90 cases of technology transfer through licensing in Latin America, then explores statistically the relationship between licensing and technological dependency. Finds a "technological dependence syndrome"--licensed transfer often prevents hands-on learning by indigenous workers.

210. Mytelka, Lynn K. 1985. "Stimulating Effective Technology Transfer: The Case of Textiles in Africa." Pp. 77-126 in *International Technology Transfer: Concepts, Measures, and Comparisons*, edited by N. Rosenberg and C. Frischtak. New York: Praeger.

Effective transfer occurs when indigenous technological capabilities are built through a conscious effort to "learn by doing" spurred by key decision makers in an enterprise, especially in the context of pressures to reduce production costs. Cases of Ivory Coast and Nigeria show how low innovativeness occurs, while Kenya and Tanzania exhibit other dynamics. State ventures are not as likely as private to engage in technological apprenticeship.

211. Natarajan, R. and Pita O. Agbese. 1989. "New Technologies, the South and Technology Transfer." *International Journal of Contemporary Sociology* 26: 25-37.

Impact of microelectronics, biotechnology and materials science developments on the economic, political, and social structures of the South. Argues they lead to increased technological dependence of the South. Covers much territory, but speculative.

212. Niosi, J. and J. Rivard. 1990. "Canadian Technology Transfer to Developing Countries Through Small and Medium-Size Enterprises." *World Development* 18: 1529-42.

Based on two samples, one that includes SMEs transferring technology to the largest or more active Third World countries and another that includes Canadian SMEs transferring technology to industrialized countries. Tests the application of theories of multinational firms developed for large corporations to small and medium-size corporations. They find that, among theories of MNCs, the industrial organization approach is probably best suited to Canadian SMEs.

This approach would suggest that Canadian SMEs had special assets, such as new products or processes, that gave them strength at home and abroad. Neither product-cycle theory nor the factor-endowment thesis is appropriate, since technology transfer took place with countries that had virtually no exports and the firms studied did not intensively use the factors of production with which Canada is endowed. All MNC theories are inadequate to explain the operations of SMEs; new theories must be developed. With regard to the appropriate technology debate, the authors find that SMEs are not, as some have suggested, alternative suppliers of technology to developing countries. The SMEs are found to be complementary rather than alternative suppliers of technology.

213. Oldham, G. 1990. "Technology Transfer to China's Offshore Oil Industry: A Case Study of International Collaboration in Policy Research." Pp. 5-22 in *Science Policy Research: Implications and Applications*, edited by D.F. Dealmeida. London: Pinter Publishers Ltd.

Reports preliminary research from an ongoing collaborative research project between the National Research Centre for Science and

Technology for Development (NRCSTD) in Beijing and the Science Policy Research Unit (SPRU). Describes the process of collaboration between the organizations, including initiation of the first project and the conduct of the research. Technology transfer to China's offshore oil industry was chosen to be the first research project because this topic enabled Chinese and British research teams to work in their own countries on opposite ends of the same project.

The first phase involved identifying the problem and dividing labor. The second lasted for two years and involved SPRU working mainly in the UK, while SPRU and NRCSTD jointly interviewed Chinese and foreign companies involved in offshore oil technology transfer. The collaboration has helped to build rapport, and demonstrated a way to carry out policy research in a developing country that is useful to policymakers. The article contains three lengthy appendices: on developing the capacities of Third World oil companies to generate and manage technological change, on experiences of the UK and Norway in building technological capacity in the offshore oil industry, and on recent experiences of the Chinese National Offshore Oil Corporation (CNOOC) in absorbing technology from foreign oil companies.

214. Otsuka, Katsuo. 1982. "The Transfer of Technology in Japan and Thailand: Sericulture and the Silk Industry." *Development and Change* 13: 421-45.

Examines the process of technology transfer in underdeveloped economies to discover ways in which this can be done successfully. The hypothesis is that technology choice is based on the principle of profit maximization in a capitalist society; the success of transfer will depend on the return to each of the factors of production. Gives the history of silk and sericulture in Japan and Thailand. Quantitative tables of return to factors of production. Similarities: (1) both countries developed dual structure after the import of modern technology, (2) both invited technicians and experts from overseas, (3) governments have played a positive role, (4) business behavior based almost wholly on principle of profit making in the long run. Dissimilarities: (1) time of adoption, (2) differences in development of traditional technology, (3) differences in the way new technology was modified, (4) differences in financing. Central issue is how to use traditional technology. Problems of technology transfer occur at early stage of development.

215. Perlmutter, Howard V. and Tagi Sagafi-Nejad. 1981. *International Technology Transfer: Guidelines, Codes and a Muffled Quadrilogue*. New York: Pergamon Press.

Studies the underlying processes that affect the ability to establish multilateral guidelines or "codes of conduct" for technology transfer. Traces history of the debate, then analyzes guidelines such as training, pricing, and arbitration using surveys of major actors in the process.

216. Poznanski, Kazimierz. 1986. "Patterns of Technology Imports: Interregional Comparison." *World Development* 14: 743-56.

Finds a heavy infusion of outside technologies into Latin America's NICs--Mexico, Brazil and Argentina--relative to that going into East European nations. Based on the assertion that Latin American countries benefit from large-scale, direct investment by Western firms, argues that Latin American countries will continue to import technology. Such action improves their attractiveness for technology-intensive goods and foreign equity investment, and might continue the progressive erosion of East European exports to the West. Offers evidence from studies of patent and licensing data.

217. Ramesh, J. and C. Weiss (eds.). 1979. *Mobilizing Technology for World Development*. New York: Praeger.

A collection of essays suggesting a new approach to the North-South technology agenda, from the Jamaica Symposium (January, 1979) by the International Institute for Environment and Development. The book is useful because of its attempts to deal with issues of development and technology in terms of global interconnections. A good, but somewhat dated, overview of the North-South dialogue on technology transfer.

218. Rath, Amitav S. (ed.). 1990. "Science and Technology: Issues from the Periphery." *World Development* 18.

Special issue devoted to science and technology. The issue is introduced by the editor's "Science, Technology, and Policy in the Periphery: A Perspective from the Centre," which offers a general discussion of science and technology issues from the perspective of the International Development Research Centre (IDRC) in Canada. Other papers deal with technical change in Latin American manufacturing firms, technology policy in sub-Saharan Africa, a multiple source of innovation model of agricultural research and promotion, a study of technological capacity and failure of industrialization in the case of Ghana's Bonsa Tyre Company, a case study of truck manufacturing in Turkey, tech transfer to developing countries by Canadian small and middle-sized enterprises, science in Latin America, S&T indicators in small countries, environmental and technological degradation in peasant agriculture in Mexico, and potentials of the software industry in Latin America.

219. Robinson, Richard D. 1988. *The International Transfer of Technology: Theory, Issues, and Practice*. Cambridge: Ballinger.

Begins with a discussion of definitional and measurement problems, then attempts to develop a general model of the transfer process. The model incorporates pressures inside and outside of firms that encourage or inhibit technology transfer. Analyzes specific issues, such as government intervention, choosing where to locate R&D, the protection of transferred technology from unauthorized use, international competitive bidding, and selection of an appropriate organizational mode for transfer. Distinguishes four organizational modes apart from the wholly-owned or majority-owned subsidiary: (1) the equity-based joint venture, (2) the contractually-based joint venture, (3) the partnership, and (4) technical collaboration by contract.

220. Robock, Stefan H. and Robert D. Calkins. 1980. *The International Technology Transfer Process*. Washington: National Academy of Sciences.

A short paper prepared for the Committee on Technology and International Economic and Trade Association to outline the myriad ways technology is transferred and indicate principal agents in the transfer process. Includes sections on measurement of technology flows and improving the database for policymakers.

221. Rosa, L.P. 1990. "Technology Transfer and Development in Brazil: Comparative Study of the Offshore Oil Case." Pp. 23-40 in *Science Policy Research: Implications and Applications*, edited by D.F. Dealmeida. London: Pinter Publishers Ltd.

A study of Brazilian policies for the development, transfer, and absorption of technology. The history of the offshore oil industry in

Brazil is described in detail, and then the offshore oil industry is compared with aeronautics, armaments, computers, and nuclear energy. Contains a useful table that compares these five areas of technology in terms of promoting Brazilian institutions (such as Petrobras, army, or Ministry of Science and Technology), principal executors, existence of previous competence in the country, formation of human resources and research, use of tech transfer versus national technology, means of absorption of technology, economic viability, and technical viability.

222. Rosenberg, Nathan and Claudio Frischtak (eds.). 1985. *International Technology Transfer: Concepts, Measures, and Comparisons*. New York: Praeger.

Good overview of a number of important authors and projects, stressing new definitions of locally specific knowledge and technology as embodied in people. Originated in a conference by the Social Science Research Council subcommittee on Science and Technology Indicators to report on measures related to ITT. Includes an overview of transfer among advanced countries, trends in North-South transfer of high technology, six chapters on individual countries (two on Japan), and a final chapter on licensing versus foreign direct investment by U.S. corporations. Country chapters contain summaries of important work by Lall, Mytelka, Katz, Westphal and Dahlman.

223. Safarian, A.F. and Gilles Y. Bertin (eds.). 1987. *Multinationals, Governments, and International Technology Transfer*. London: Croom Helm.

A collection of essays from a conference in Paris by French and Canadian scholars on the role of MNCs in technology transfer. Although the case studies center on industrialized nations, many essays are more general in nature about the problems involved in transfer. Focus on strategies of MNCs, particularly to what extent transfers are internal, and how successful they are from the viewpoint of both the firm and the countries involved.

224. Salas Capriles, Roberto. 1977. "Technology Transfer and the Industrialists in Latin America." *Impact of Science on Society* 27: 307-20.

Sees technology as a commodity, and the transfer of technology as a process in which a country is free to choose autonomously from different alternatives. The basic question is: how can Latin America develop its own technology? Stresses the need for entrepreneurship.

225. Samli, A.C. (ed.). 1985. *Technology Transfer*. London: Quorum Books.

Chapter 1 ("Technology Transfer: The General Model") covers Dimensions of technology transfer: geography, culture, economy, people, business, and government. Chapter 2 ("Technology Transfer to Third World Countries and Economic Development") argues that technology transfer is the best strategy for LDC development. LDCs commonly suffer from a vicious cycle of underdevelopment. Four strategies can help break the cycle: (1) increased savings, (2) increased international trade, (3) foreign aid, and (4) technology transfer. But the first three create such significant problems that TT becomes most desirable. TT can break the vicious cycle by enabling the country to substitute imports, or just by improving the country's overall economic efficiency.

226. Savio, Roberto. 1987. "TIPS: An Information System for South-South Communication." *Development Issues* 2&3: 135-41.

Outlines a communication system intended to facilitate technology transfer between third World nations.

227. Schnepp, Otto, Mary Ann von Glinow, and Arvind Bhambri. 1990. *United States-China Technology Transfer*. Englewood Cliffs, NJ: Prentice Hall.

Two basic questions: (1) What are the critical issues confronted by U.S. companies transferring technology to China and by Chinese companies acquiring technology from the United States? (2) What are effective and ineffective strategies for managing these issues? To this end, the authors offer eight case studies of technology transfer. These are four technology transfer situations (Shanghai-Foxboro Co. Ltd., Westinghouse Electric Corp., Cummins Engine Co., and Combustion Engineering) examined first from the U.S. and then from the Chinese perspective. The final part of the book is devoted to a comparative evaluation, consisting of a comparative assessment and implications for management. The introduction of the book contains a good summary of current trends in research literature on technology transfer.

228. Segal, Aaron. 1987. *Learning by Doing: Science and Technology in the Developing World.* Boulder, CO: Westview Press.

Looks at the difficulties involved with moving from technology importation to technology self-sufficiency, with chapters covering Latin America, the Caribbean, the Middle East, Africa, East Asia, China, and India. Each chapter outlines the historical experiences of these regions and projects likely to succeed in the future.

229. Seitz, Frederick. 1982. "The Role of Universities in the Transnational Interchange of Science and Technology for Development." *Technology In Society* 4: 33-40.

Emphasizes cultural factors influencing transfer of technologies. In particular, universities can promote interchange by educating professionals to absorb technology, adding to their "cultural receptivity" of innovations. Also, universities help by housing the scientific community in institutions promoting technological cooperation.

230. Servaes, Jan. 1990. "Technology Transfer in Thailand: For Whom and For What?" *Journal of Contemporary Asia* 20: 277-87.

Address given at a colloquium on technology transfer that exhibits Third World "fear of dependency" and awareness of the need for balance between self-reliance and strategic imports. Thailand is rapidly resembling a colony, as an export-oriented economy which offers all kinds of facilities to transnationals. Western technology can't be borrowed without taking in culture at the same time. Three implicit values in Western technology: (1) little respect for myth, symbol, or the power of the mysterious, (2) technology is based on cult of efficiency, (3) technology dominates and manipulates nature rather than being in harmony with it. Review of three different views of technology transfer (modernization, world economy, ecological).

231. Shaikh, Rashid A. 1986. "The Dilemmas of Advanced Technology for the Third World." *Technology Review* 89: 56-64.

Broad examination of the problem of hazardous materials in advanced technology, dealing with the export of these materials and with pollution by American overseas factories. Suggests that uncritical attempts to modernize can be dangerous and that countries and international lending agencies should look at all forms of impact before adopting a development strategy.

232. Sharif, M.N. and A.K.M.A. Haq. 1980. "Evaluating the Potentials of Technical Cooperation Among Developing Countries." ***Technological Forecasting and Social Change*** **16: 3-32.**

Argues that an LDC's "optimal partner" for technology transfer might be another LDC, and presents a method for determining the optimal partner using computerization technology as an example.

233. Shayo, L.K. 1986. "The Transfer of Science and Technology Between Developed and Developing Countries Through Co-Operation Among Institutions of Higher Learning." ***Higher Education in Europe*** **11: 19-23.**

Outlines five methods of transferring science and technology to developing countries: (1) voluntary expatriates, (2) personnel assistance, (3) personnel training, (4) technology importation, and (5) cooperation among institutions of higher learning. Focuses on the fifth kind through example of a Middle Eastern university.

234. Siddarthan, N.S. 1988. "In-house R&D, Imported Technology, and Firm Size: Lessons from Indian Experience." ***Developing Economies*** **26: 212-21.**

Looks at the interrelationships among size of firm, R&D, and imports. Concludes that the relationship between import of technology and domestic R&D expenditures is complementary only for private-sector firms, and that this complementarity is stronger for low-tech imports. Relationship of firm size to R&D expenditure is nonlinear, since R&D expenditure increases more slowly than firm size for smaller enterprises and faster than firm size for larger enterprises.

235. Sietz, F. 1982. "The Role of Universities in the Trans-national Interchange of Science and Technology for Development." ***Technology in Society*** **4: 33-40.**

Outlines the university experience in technology transfer for a number of nations or regions--Scandinavia, the U.S. (Marshall Plan), the UK, Japan, Taiwan, and South Korea. Then offers account of a new U.S. cooperative program between the U.S. National Academy of

Science and the Agency for International Development (AID).

236. Siggel, Eckhard. 1983. "The Mechanisms, Efficiency and Cost of Technology Transfers in the Industrial Sector of Zaire." *Development and Change* 14: 83-114.

Suggests that the most important way to break dependence is to train workers, not terminate technology transfer. In the first section of the article, a learning-oriented concept of technology transfer is developed. In the second section, the principal transfer mechanisms are described and analyzed with respect to kinds of technology transferred, using evidence from Zaire gathered in a field study of 13 manufacturing industries scattered about the country. In this study, over 200 managers in 63 enterprises were interviewed and several plants were studied in depth. The third section examines channels of technology for effectiveness, appropriateness of technology transferred, and cost of transfers.

237. Simon, Denis Fred. 1986. "The Challenge of Modernizing Industrial Technology in China: Implications for Sino-U.S. Relations." *Asian Survey* 26: 420-39.

Focusing on the program for technical transformation of enterprises, the article highlights bottlenecks and problems with Chinese management of technology transfer and communication. Primary areas of interest are biotechnology, microelectronics, information technology, and materials. Problems in technology transfer: tendency to seek out only the most advanced technology, overemphasis on hardware, lack of clarity with respect to overall imports, slow decision making.

238. Simon, Denis Fred. 1991. "International Business and the Transnational Movement of Technology: A Dialectic Perspective." Pp. 5-28 in *Technology Transfer in International Business*, edited by Tamir Agmon and Mary Ann von Glinow. New York and Oxford: Oxford University Press.

Looks at facilitators of and barriers to expanded technology transfer. Most important among the facilitators is the evolution of a shorter product life cycle. A second factor has been the convergence of international capital and factor markets. The third facilitator has been advances in international communication and transportation. "Techno-nationalism" is identified as the most important barrier to expanded tech transfer. Yet new forms of collaboration are emerging. The rapid movement of technology overseas contributes to the strategic alliance as a new form of international business venture. These new forms generally emphasize bilateral transfers of technology, while traditional forms of technology transfer tend to emphasize unilateral flows.

239. Simon, Denis Fred and Merle Goldman (eds.). 1989. *Science and Technology in Post-Mao China*. Cambridge: Harvard University Press.

Contains four categories of essays: Historical Precedents, The Reorganization of Science and Technology, Application of the Science and Technology Reforms, and Technology Transfer. An introduction by the authors provides a political and cultural context for the essays. Primarily a historical work, with good insights on the role of science in a centrally planned economy.

240. Singer, H.W. 1988. "Transfer of Technology: A One-Way Street." Pp. 3-16 in *Technology Transfer By Multinationals*, edited by H.W. Singer, Neelambar Hatti, and Rameshwar Tandon. New Delhi: Ashish Publishing House.

Simple overview of the conditions under which technology transfer occurs, and its appropriateness. Discusses patent and trademark systems, international cooperation.

241. Singer, H.W., Neelamar Hatti, and Rameshwar Tandon (eds.). 1988. *Technology Transfer by Multinationals*. New Delhi: Ashish Publishing House.

Two volume set with sections on issues of conflict and control, global efforts for tech transfer, MNCs, technological development in host countries, technological change in the North and options for the South, the international patent system, a code of conduct for MNCs, and technological interdependence.

242. Soete, Luc. 1985. "International Diffusion of Technology, Industrial Development and Technological Leapfrogging." *World Development* 13: 409-22.

How the international diffusion of technology can be understood in terms of a historical process of industrialization spurts.

Offers a review of important contributions to the technology diffusion literature and discusses the relevance of these contributions to structural theories of industrial development of economic growth. Discusses the issue of international diffusion, with a focus on historical changes in technological leadership. Argues that the potential for developing countries to leapfrog into microelectronics has never been higher than at the present.

243. Solo, Robert A. 1975. *Organizing Science for Technology Transfer in Economic Development*. Lansing: Michigan State University Press.

A study of how Britain, Germany, France and the Netherlands organized scientific resources within and outside OECD programs to promote economic development in underdeveloped countries during the late 1960s and early 1970s.

244. Starr, Paul D. 1982. "International Barriers to the Transfer of Agricultural Technology: A View from the Land-Grant Universities of the USA." *International Review of Education* 28: 485-8.

Discusses the role of land-grant universities in transferring technology as a result of the 1975 Title XII Amendment to the Foreign Assistance Act. Suggests that institutional and attitudinal problems prevent these universities from acting as effectively as possible.

245. Stewart, Charles T. and Yasumitsu Nihei. 1987. *Technology Transfer and Human Factors*. Lexington, MA: Lexington Books.

Looks at human resource development as a means of technology transfer. General education and technology-specific education are two broad types of such development needed to transfer technology. Absorptive capacity for transfer is seen as composed of the educational system and the infrastructure. Examines the local environment for technology transfer in Thailand and Indonesia, looks at American and Japanese contributions to transfer by means of human resource development in these two countries, and compares the contributions of the U.S. and Japan. Based on existing situations in Thailand and Indonesia, the authors make recommendations on what kind of contributions are still needed, recommendations to U.S. and Japanese firms and business organizations, and recommendations to

U.S. and Japanese governments. Suggests some of the areas of human resource development in which Japan and the U.S. can cooperate, and what the roles of Thailand and Indonesia in their own resource development should be.

246. Stewart, Frances. 1988. "Technology: Major Issues for Policy in the 1980s." Pp. 17-45 in *Technology Transfer By Multinationals*, edited by H.W. Singer, Neelambar Hatti, and Rameshwar Tandon. New Delhi: Ashish Publishing House.

Good general review of main issues in North-South technology transfer, including choice of technology, terms of transfer, the build-up of tech capabilities in LDCs, and the implications of micro-electronics. Sees the need for LDCs to adopt a selective policy for both production and consumption, a strong negotiating stance, local technological capability to improve efficiency of use, R&D on adaption, and South-South cooperation.

247. Stewart, Frances. 1990. "Technology Transfer for Development." Pp. 301-324 in *Science and Technology: Lessons for Development Policy*, edited by Robert E. Evenson and Gustav Ranis. Boulder & San Francisco: Westview Press.

Considers several questions: (1) What policies will help governments maximize benefits and minimize costs of technology transfer? (2) What forms of transfer should be encouraged? (3) How much should be transferred (in which industries and at what price)? (4) What should government policy be toward R&D and other aspects of the technological infrastructure? and (5) How do general economic policies influence benefits and costs of technology transfer? Concludes that developing country governments must attempt to strike a balance between the increases in productivity that modern technologies make possible and the high costs of oligopolistic supporters, as well as the inappropriateness of much imported technology and import dependence. Appropriate tech strategies depend on stage of development. The least developed must import, but they can minimize the price by sticking to older technologies, by searching for Third World suppliers, and by developing their own human capacities. Middle stage countries are in a better position to exploit new technologies. Technological advances are creating a division that makes increases the difficulty for newcomers trying to repeat the success of the NICs.

248. Subrahmanian, K.K. and P. Mohanan Pillai. 1988. "Technology Transfer: Critical Issues and Strategy Options--A Review of the Indian Situation." Pp. 307-335 in *Technology Transfer By Multinationals*, edited by H.W. Singer, Neelambar Hatti, and Rameshwar Tandon. New Delhi: Ashish Publishing House.

Examines Indian strategies in technology transfer. Includes empirical research on terms of transfer and costs of imports, the characteristics of imported technology, and learning effects.

249. Tatkin, Ho. 1985. "The Japan-Singapore Institute of Software Technology: A Case Study in Technology Transfer." *Education and Computing* 1: 249-63.

Traces the development of a training institute in which computer software technology is transferred from Japan to Singapore. Rather than use a traditional approach to training, the institute has simulated a working environment. By this approach, with a combination of Japanese and local lecturers, the Institute has succeeded in training a pool of analyst/programmers and systems analysts.

250. Teece, David J. 1976. *The Multinational Corporation and the Resource Costs of International Technology Transfer.* Cambridge, MA: Ballinger.

Examines the costs of international technology transfer with data from U.S.-based MNCs on 29 projects. Projects selected predominantly from the chemicals industry; some of these involved transfers to developing countries, but the special problems here are not stressed. Offers hypotheses on level and determinants of cost.

251. Teece, David J. 1981. "The Market for Know-How and the Efficient International Transfer of Technology." *Annals of the American Academy of Political and Social Science* 458: 81-96.

Argues that transfer resource costs depend upon the transmitter, the receiver, the technology being transferred, and the institutional mode chosen for transfer. Discusses relation between the codification of knowledge and costs of transfer: uncodified or tacit knowledge is slow and costly to transmit. Technology transfer by either importers or exporters can't substantially improve the efficiency with which this market operates.

252. Teitel, Simon and Francisco C. Sercovich. 1984. "Latin America." *World Development* 12: 645-660.

Developing countries began to export technology during the 1970s. Presents information from country studies on Argentina, Brazil, and Mexico, and preliminary reports on the Andean subregion and on technology purchasers. Examines (1) the kinds of technological goods and services that are being exported and kinds of export channels being used, (2) the destination of the exports, (3) whether the exporting agents are public, private, multinational or local exporters, (4) institutions and policies related to technology exports. Also conducts an evaluation of the state of Latin American technology exports and its prospects.

Finds that many technology exports have taken place and that the value of construction contracts is higher than that of industrial technology exports. The exports represent a substantial part of the foreign technology accounts of these countries and most of the exports, especially those for industrial projects, are made within the region. Government policies of openness to international trade and support in finding new markets play an important role in promoting exports. The establishment of local consulting firms and personnel training has an indirect effect. Exports of construction services are closely and inversely related to the domestic investment cycle. At the sectoral level, promotion of a specific activity or industry may be important.

253. Thomasemeagwali, G. 1991. "Technology Transfer: Explaining the Japanese Success Story." *Journal of Contemporary Asia* 21: 504-512.

Looks at technology transfer in Japan with a view to circumstances that have facilitated Japanese economic and technological success. Provides historical background of the transformation of Japan from a precapitalist to an advanced industrialized nation. Identifies local and systemic developments that were conducive to the diffusion of technology. Japan adopted industrial technology at the turn of the century, when technology was changing rapidly. Japan was able to profit from intra-European rivalries at a crucial period of technology transfer, and benefitted from being a junior partner of the U.S. in the postwar period.

254. Tuma, Elias H. 1987. "Technology Transfer and Economic

Development: Lesson in History." ***Journal of Developing Areas*** **21: 403-28.**

Presents 5 propositions: (1) technological advancement has been the most important source of economic development and growth, (2) technology has been transferred from one economy to another but the cost has sometimes been prohibitive, (3) cost of technology transfer is a function of the development gap between the economies across which the transfer occurs, (4) transfer has come about more frequently as a result of planning and intervention in the market than in response to invisible market forces, (5) developing countries have suffered because they have adopted a mutilated concept of transfer and have sustained a defective context.

Gives an historical examination of economic development and technological revolutions in developed countries. Changes in technology are associated with Kondratiev cycles. Conditions responsible for the transfer of technology in developing countries are availability of know-how and existence of a capital goods industry. Multinational corporations obstruct domestication of know-how and perpetuate dependence.

If forces of supply and forces of demand favor diffusion of technology and assimilation into domestic economy, why has the low level of technology in the Middle East persisted? Problems in education, a confused relation with religion, conflict with the traditions of the region, inadequate support for experts from mid-level technicians and vacillating political leadership. Middle Eastern countries have no choice but to transform technologies to higher levels if they are to develop their economies.

255. Varas, Augusto and Fernando Bustamante. 1983. "The Effect of R&D on the Transfer of Military Technology to the Third World." ***International Social Science Journal*** **35 : 141-61.**

R&D made the arms industry profitable. As long as the U.S., France, and the UK maintained dominance and control of markets, no one resorted to arms sales to improve trade deficits. In the 1960s Germany and Japan had civilian industries only. Manufacture under license is one of the most important forms of military technology transfer to the Third World. Discusses socio-political effects and economic effects of the transfer.

256. Vayrynen, Raimo. 1978. "International Patenting as a Means of Technological Dominance." *International Social Science Journal* 30: 315-38.

Maintains that the high degree of monopolization in technology markets has allowed those controlling technology to use patents, licenses, and trademarks as devices by which to regulate the international flow of technology. Therefore, patents prevent rather than promote the transfer of technology to peripheral countries.

257. Vickery, Graham. 1986. "International Flows of Technology: Recent Trends and Developments." *STI Review* 1: 47-83.

Examines three routes for the transfer between countries of new and improved products and production methods: (1) licensing or sale of patents, intellectual property rights or know-how; (2) R&D performed by foreign firms; and (3) foreign direct investment. Finds that most international trade in patents and know-how takes place among OECD countries, that technology trade is dominated by MNCs, and that R&D is internationalizing rapidly to take advantage of local technological resources and partners. The structure of foreign direct investment has changed markedly, with the U.S. becoming the largest destination for foreign direct investment flows.

258. Wang, Jian-Ye. 1990. "Growth, Technology Transfer, and the Long-Run Theory of International Capital Movements." *Journal of International Economics* 29: 255-71.

Mathematical economics. Provides a model of growth and international capital movements. Technology is assumed to be transferred from the developed North to the developing South. When South shifts to free capital mobility, its steady-state growth rate of per capita income rises. Under free capital movements, an increase in the autarkic rate of human capital and/or technological diffusion rate in the South lowers the steady-state income gap between the North and South. Policy implications: (1) LDCs profit from opening to direct foreign investment and (2) an initially backward country may reduce the per capita income gap between itself and more advanced countries.

259. Washington, R.O. 1984. "Designing a Management Information System in the Arab Republic of Egypt: A Case Study of Factors Influencing Technology Transfer in Third World

Countries." *Social Development Issues* 8: 158-71.

Argues that institutions and "value climates" of Third World countries affect the way new knowledge is received and used, and therefore affects the success of technology transfer. Qualitative data to illustrate this thesis are drawn from the author's personal experiences and observations as a result of his involvement in the design and implementation of an information and reporting system for the Ministry of Social Affairs in Egypt. The author concludes that development workers need to be trained in the history of their host countries, and that these workers should be skilled in the use of "ethnomethodological cues."

260. Weiss, Bernard and Thomas W. Clarkson. 1986. "Toxic Chemical Disasters and the Implications of Bhopal for Technology Transfer." *The Milbank Quarterly* 64: 216-40.

Argues that, in addition to measures to prevent sudden and drastic occurrences like the accident at Bhopal, the problem of delayed toxicity also should be considered in the process of technology transfer. Gives examples of gradual toxicity in developed and developing countries. Maintains that, as developing countries take up the technology of developed countries, they also must take up such toxicological perspectives and practices as early biological monitoring of those exposed and extrapolation from animal data.

261. Williams, Frederick and David V. Gibson. *Technology Transfer: A Communication Perspective.* Newbury Park, CT: Sage Publishers.

A collection of essays on: (1) challenges of technology innovation and transfer, (2) the organizational setting of technology transfer, (3) contexts of technology transfer, (4) international perspective, (5) the literature of technology transfer. The fourth section includes a discussion of Mexico, India, and MNCs.

262. Williams, John Hoyt. 1977. "Foreign Tecnicos and the Modernization of Paraguay, 1840-70." *Journal of Interamerican Studies and World Affairs* 19: 233-57.

Historical study of Paraguay's use of foreign experts during the rules of Carlos Antonio Lopez and Francisco Solano Lopez. Finds that Paraguay achieved rapid modernization through relying on imported skills, rather than imported capital. This enabled Paraguay to escape the pressures brought to bear on other Latin American countries by European investors. As a result, by 1864 Paraguay was debt-free and technologically advanced by comparison to other Latin American countries. However, all the modernization in this period was military or defensive in nature. Paraguay's advantages were destroyed by war in the 1860s.

263. Wionczek, Miguel S. 1976. "Notes on Technology Transfer through Multi-national Enterprises in Latin America." ***Development and Change*** **7: 135-55.**

Examines evidence for the hypotheses stated in an OECD paper by Prof. C.A. Michalet. (1) MNEs play an essential role in dissemination of technology (large percent of sales consists of internal tech trade between MNEs and subsidiaries), (2) low-tech industries rank higher in royalties than high tech--when high tech is concerned, transfer is accomplished by export of new products rather than direct investment, (3) within MNEs, the book price and the real price of any internal technology transfer do not correspond, (4) MNEs devote more resources to R&D and to innovation used in new products than to introduction of new production processes, (5) patterns of R&D and tech transfer involve conflict between MNEs and host countries, (6) R&D activities of MNEs are highly centralized in parent companies, (7) R&D activities of subsidiaries are highly controlled by parent companies, (8) the R&D burden is unevenly distributed, to the detriment of the subsidiary, as a result of the product cycle, (9) global strategy of MNEs is a factor generating dependence on the part of recipients.

264. Wionczek, Miguel S. 1986. "Industrialization, Foreign Capital and Technology Transfer: The Mexican Experience 1930-85." ***Development and Change*** **17: 283-302.**

Historical study of the development of Mexico, showing heavy costs to the agricultural sector. Costs are increasing, and in areas where the agricultural sector can hold off the costs, development is still difficult.

265. Young, G.B. 1985. "The Transference of Technology to, from, and within South East Asia." ***Asian Affairs*** **16: 5-19.**

Historical view of global transfer. Any transfer of technology almost invariably follows in response to some demand, which can be influenced not only by commercial or economic factors but also by social factors. Technology divided into three categories--high, intermediate, low. Transfer takes place (1) through outright sale of goods incorporating technology, (2) by owner entering into joint venture, (3) by owner transferring to wholly owned operation in another country. From the standpoint of recipient, these considerations require attention: (1) Is tech at right level for country concerned? Tech should satisfy criteria of (a) cost, (b) suitability for sophistication of workforce, (c) country's requirement to create or reduce employment opportunities; (2) best value overall, not just initial, cost. If acquired for export, final product not be obsolete.

266. Zahlan, A.B. 1978. *Technology Transfer and Change in the Arab World*. Oxford: Pergamon.

A series of seminar papers addressing questions of Arab tech transfer, with articles on patterns of acquisition, the role of government policies, the behavior of international organizations, the problems of indigenous innovation, and the importance of values. Several essays focus on individual economic sectors: water use, agriculture, transportation, food, solar energy, mineral resources, and steel.

Chapter 4 -- Economics and Manufacturing

267. Amsalem, Michel A. 1983. *Technology Choice in Developing Countries: The Textile and Pulp and Paper Industries*. Cambridge, MA: MIT Press.

Excellent study based on a survey of pulp/paper, and textile equipment suppliers and analysis of the technology choices made for production facilities by 28 firms in developing countries (Colombia, Brazil, Philippines, Indonesia). Detailed description of production processes, alternative technologies, and measurement strategies. Separate chapters on production cost minimization and technology choices for each of the two industries. Main questions: how wide is the choice of technologies and is there much variation in the proportions of factors of production? What is the role of cost minimization and are there other factors which influence choice? In most cases, the technologies chosen are not those that would lead to the lowest production costs for the firm. Important influences are availability and cost of information, minimization of risk, and government policies.

268. Amsden, A. 1977. "The Division of Labour is Limited by the Type of Market: The Case of the Taiwanese Machine Tool Industry." *World Development* 5: 217-34.

Maintains that the division of labor is limited by the type of market (the average income of individual consumers), as well as by the size of the market (aggregate purchasing power). Therefore, capital accumulation is needed for technology to advance in production of certain goods. Multiplication of markets is not sufficient by itself. Examines the Taiwan machine tool industry in order to provide empirical support for this proposition.

269. Antle, John M. and Charles C. Crissman. 1988. "The Market for Innovations and Short-Run Technological Change: Evidence from Egypt." *Economic Development and Cultural Change* 36: 669-90.

Derives from Egyptian evidence the conclusion that the market for innovations need not reach an equilibrium to induce technological change. When factory prices deviate from long-term trends, farmers are more likely to perceive the change as permanent and alter expectations while researchers continue to supply innovations. Short-run deviations in real wages do not appear to affect technology development; in particular, labor-intensive technologies continue to be produced despite wage fluctuations. This disequilibrium is seen as suboptimal because labor-saving innovations are not adopted when the market would so dictate.

270. Ayres, Robert U. 1985. "Social Technology and Economic Development." *Technological Forecasting and Social Change* 28: 141-57.

Asks why Third World nations do not have the best technologies available, pointing to social causes and focusing on population growth, resources, environmental issues, and economic development. Basic problem of economic development is not scarcity of resources or excessive technological dependence, but resistance to change and lack of innovation.

271. Baer, Werner. 1976. "Technology, Employment and Development: Empirical Findings." *World Development* 4: 121-30.

A review of recent books by A.S. Bhalla and A. Sen on low-level labor absorption in Third World industrial sectors. Praises Sen's work for providing a lucid, critical summation of current theoretical knowledge, and Bhalla's work for supplying empirical case studies using cross-section, time series and input-output analyses that support Sen's generalizations. The author disagrees with the optimistic view that government and private sector businesses can and ought to influence the degree of labor absorption in LDCs, claiming such a solution would condemn developing countries to stagnation. Particularly emphasizes the role of service sector employment.

272. Bagachawa, M.S.D. 1992. "Choice of Technology in Small and Large Firms." *World Development* 20: 97-108.

Evaluates the relative performance of small- and large-scale grain milling techniques in Tanzania to identify appropriate techniques and to explain why some firms select inappropriate techniques and products. Data are taken from a cross-sectional survey of 49 maize and 16 rice milling units scattered around the country. Secondary information comes from government ministries and institutions. Appropriate technology is defined as technology in accord with Tanzania's basic industrial strategy (BIS) which emphasizes the use of domestic resources to produce essential commodities for consumers. Finds that small-scale custom milling is economically viable in terms of employment generation and effective utilization of capital and surplus generation, and that the custom milling subsector provides vital milling services to the majority of the low-income population in both rural areas and small towns. The low initial investment costs and the operational manageability of the mills provide additional incentive for development of the small-scale entrepreneurial sector.

273. Bell, Martin, B. Ross-Larson, and L.E. Westphal. 1984. "Assessing the Performance of Infant Industries." ***Journal of Development Economics*** **16: 101-28.**

Reports findings from a literature survey to learn about the performance of infant industries in less-developed countries. Little direct evidence was uncovered about the costs and benefits of developing these industries. Findings indicate that infant firms have experienced relatively slow productivity growth. This insufficient productivity growth, which inhibits the ability to achieve and maintain competitiveness, appears to be due to the absence of sustained efforts to acquire and use the capabilities necessary for continuous technological change.

274. Bhagwati, Jagdish N. 1979. "International Migration of the Highly Skilled: Economics, Ethics and Taxes." ***Third World Quarterly*** **1(3): 17-30.**

Discussion of the social and economic consequences and causes of "brain drain."

275. Bhalla, A.S. (ed.). 1981. ***Technology and Employment in Industry*****. Geneva: ILO.**

Part 1 deals with conceptual and measurement issues, including the concept and measurement of labor intensity, a review of empirical evidence on capital-labor substitution possibilities, and a discussion of indirect employment effects of investment. Part 2 consists of 10 empirical case studies. The last chapter attempts to draw together the main inferences from the case studies and brings out key issues and implications for policy-making. Among the overall conclusions are the following: substitution possibilities exist in both core and ancillary operations; the range of available techniques can be widened by redesigning or copying older designs and blueprints with local engineering and adaptations, or through local manufacture of equipment; use of capital-intensive techniques rather than equally efficient labor-intensive ones is due to imperfect knowledge and inappropriate selection systems.

276. Blumenthal, W. Michael. 1988. "The World Economy and Technological Change." ***Foreign Affairs*** **66: 529-550.**

Author is chairman and CEO of UNISYS. Argues that many of the problems the U.S. faces are a result of the "extraordinarily rapid technological change" of the 1970s and 1980s, due to the lag between the rate of technological change and the rate of adjustment to changes among decision makers. Three conclusions regarding technological change: (1) technology is making the basic notion of national sovereignty obsolete in many areas of economic affairs, (2) nation-states will continue to exist and will behave as if they can control key events more effectively than they actually can, and (3) an integrated view of new issues underlying current economic problems in the light of technological changes is needed.

Suggests that even with the correct understanding of these problems, a constructive response can occur only with the leadership of the U.S. This response should take the form of a general framework for the gradual restructuring of economic relations that corresponds more closely to technological realities. This framework needs five principles: (1) new strategy should be based on willingness to think about the world as it is and not as it was, (2) U.S. hegemony will be replaced by the triangular bloc of the U.S., Japan, and EEC, (3) the concept of "national interest" needs to be redefined to limit unilateral action in economic affairs, (4) measures incompatible with an expanded definition of national interest should be renounced, and (5) broader interdependencies with respect to economic problems should be taken into account.

277. Boon, Gerard K. 1978. ***Technology and Sector Choice in Economic Development.*** **The Hague: Sijthoff & Noordhoff.**

Uses various economic methodologies to assess production

technologies and economic sectors relative to their appropriateness for economic development. Specific analyses emphasize Mexico; the final third of the book is largely a discussion of the political economy of Mexico. The first section uses microanalysis to discuss the "DOS" method for evaluating and selecting technology. One section emphasizes sectoral analysis.

278. Bowonder, B. and T. Mijake. 1988. "Measuring Innovativeness of an Industry: An Analysis of the Electronics Industry in India, Japan and Korea." *Science and Public Policy* 15: 279-303.

Offers a new approach for measuring innovativeness which consists of: measuring characteristics of technology, estimating export competitiveness of outputs, and assessing the output by phases of technology life cycle. Finds the following prerequisites for technology development: specialization in manufacturing, commitment for technological upgrading as well as assimilation, technology consideration within economic planning, and the climate conducive for such development.

279. Burstein, M.L. 1984. "Diffusion of Knowledge-Based Products: Applications to Developing Economies." *Economic Inquiry* 22: 612-33.

Because of the free rider problem, an innovator often finds initial product development plans unproductive and must resort to alternatives. Feasible diffusion requires that the product space allocated to an innovator by grants of property rights be larger than one confined to primary inventions.

280. Chen, M.L. 1991. "The Role of R&D Subsidies When Incomplete Information Is an Entry Barrier." *Journal of International Economics* 31: 251-70.

Explores the role of R&D subsidies in the presence of incomplete information. Considers a market for experience goods where a firm's R&D investment determines the quality of its product but where quality becomes observable to consumers only through experience with the product. Externalities emerge in this model because individual firms cannot fully appropriate the benefits of quality-enhancement outlays. Demonstrates that R&D subsidies can internalize this externality.

281. Chudnovsky, D., M. Nagao, and S. Jacobsson. 1984. *Capital Goods Production in the Third World: An Economic Study of Technical Acquisition*. London: Frances Pinter.

Looks at problems, both experienced and anticipated, in the process of development of the capital goods sector in the Third World, focusing on the technological components. The first chapter examines the technological requirements involved in the design and manufacture of capital goods, on the mode of production, and on the industrial structure in which manufacture has taken place. The second chapter discusses technology problems faced by the majority of developing countries, relying mainly on UNCTAD case studies in Tanzania, Tunisia, Thailand, and Peru. The third chapter gives a detailed discussion of technology issues relating to the entry into design and manufacture of complex capital goods by the more advanced developing countries. Chapter 4 looks at the experience of China in the development of its capital goods industries. Chapter 5 examines the production of computer controlled lathes in developing countries.

282. Cimoli, M. and Giovanni Dosi. 1988. "Technology and Development: Some Implications of Recent Advances in the Economics of Innovation for the Process of Development." Pp. 117-147 in *Science, Technology and Development*, edited by Atul Wad. Boulder: Westview Press.

Discusses recent advances in the economics of innovation and technical change and analyzes their implications for economic development. Provides a summary of the major characteristics of technology and technical change, suggesting hypotheses on the dynamics of technological accumulation and on the linkages between industry-specific and country-wide processes of technological learning. Whether countries catch up, fall behind, or leapfrog depends on dynamics of the technological domain, together with institutional forms of social organization.

283. Columbo, U. 1991. "The Technological Revolution and the Future of the Third World." *Technology and Society Magazine* 10: 25-32.

A speech presented at the Munich IEEE meeting held October,

1988. Non-technical overview that argues the Third World is about to enter a technological revolution unlike the waves of innovation in developed countries over the past 50 years. This view is based on a Schumpeterian interpretation of Kondratiev waves (economic waves of boom and bust, partially caused by clusters of new technologies). The author reviews various social implications of the new technologies for the developing world and is optimistic about their ultimate impact.

284. Cooper, Charles. 1988. "Supply and Demand Factors in Indian Technology Imports: A Case Study." Pp. 105-135 in *Technology Absorption in Indian Industry*, edited by Ashok V. Desai. New Delhi: Wiley Eastern Limited.

Empirical study of Indian technology trade focusing on the Benelux countries. Concludes that liberalization of Indian policy toward technology collaboration could result in expansion of demand for European technology. Appendices contain tables showing influence of Benelux supply conditions on Indian technological collaborations and time pattern of Benelux collaborations with India.

285. Dahlman, Carl J. and Francisco C. Sercovich. 1984. "Exports of Technology from Semi-Industrial Economies and Local Technological Development." *Journal of Development Economics* 16: 63-100.

Analyzes technological exports from five relatively advanced Third World countries, and looks at local advantages such as technological development that aided the growth of these industries.

286. Deardorff, A.V. 1990. "Should Patent Protection Be Extended to All Developing Countries." *World Economy* 13: 497-508.

Analyzes patent protection based on a simple model of how patent rights affect innovation and market structure. Considers whether GATT should be involved in protection of intellectual property and suggests that it should not. Since the mission of GATT is to promote, not restrict, the free flow of goods, it should also favor the free flow of ideas. Asks whether intellectual property rights are really rights and argues, first, that the Western conception of intellectual property rights is culture-specific; and, second, that even in Western culture protection is extended to only certain forms of intellectual property. Examines the costs and benefits of extending patent protection to intellectual property.

Argues that some portion of the world should be exempt from a system of patent protection, in order to maximize the welfare of the world as a whole.

287. DeGregori, Thomas R. 1978. "Technology and Economic Dependency: An Institutional Assessment." *Journal of Economic Issues* 12: 467-76.

An institutional (Veblen/Ayres) view of technology and economic dependency. Institutionalism views technology as the dynamic force for social change, while traditional social beliefs and practices are considered the forces that resist change. Maintains that all are technologically interdependent.

288. Del Campo, Enrique Martin. 1989. "Technology and the World Economy: The Case of the American Hemisphere." *Technological Forecasting and Social Change* 35: 351-64.

Because Latin America is perceived as a unified region, it enjoys the capacity to engage in independent negotiations and coordinates international actions. This independence is crucial in a world where international influences, such as the debt crisis, have such dramatic effect on internal national economies and technological differences.

289. Desai. Ashok V. 1988. "Technological Performance in Indian Industry: The Influence of Market Structures and Policies." Pp. 1-29 in *Technology Absorption in Indian Industry*, edited by Ashok V. Desai. New Delhi: Wiley Eastern Limited.

Gives historical sketches of Indian market structures and governmental policies regarding import of technology. Shows how these have affected technological performance. Suggests that the major objectives of technology import policy have been: (1) foreign exchange conservation, (2) regulation of competition, and (3) self-reliance. Concludes that, in contrast to goods, import restrictions on technology do not produce import-replacing technology, that official restrictions affect the quality of technology, that greater government neutrality between state-owned and private firms is necessary to fully exploit imported technology, and that emphasis needs to be shifted from regulating import of technology to encouraging diffusion.

290. Dollar, D. 1986. "Technological Innovation, Capital Mobility and the Product Cycle in North-South Trade." ***American Economic Review*** **76: 177-90.**

Constructs a dynamic general equilibrium model of North-South trade that attempts to combine the product cycle approach with pressures toward factor-price equalization in the neoclassical trade model. The main insight is that for factor prices and terms of trade to be stable, a stable ratio of the number of goods produced in each region is needed. If the North is constantly introducing new products, this ratio can only be stable if technology flows to the South. Labor must earn a greater reward in the North than in the South, incentive will exist for this flow. This means that, without constant innovation, the transfer of capital and technology to the South would undermine Northern wages.

291. Dore, Ronald. 1989. "Technology in a World of National Frontiers." ***World Development*** **17: 1665-76.**

Maintains that nationalism and the intensification of international competition are the principal accelerators of scientific knowledge. In contrast, earlier in the century the institutionalization of research was the principal accelerator. Marketplace competition between commercial firms has been supplanted by competition among nations. Supports this argument with a variety of examples, such as American concern with Japanese competition in the semiconductor industry. This economic rivalry is not viewed by the author as an ideal state of affairs, but preferable to military or prestige rivalry.

292. Dunning, J.H. 1988. ***Multinationals, Technology, and Competitiveness.*** **London: Unwin Hyman.**

Uses the author's "eclectic paradigm of international production" to look at aspects of the relationship between technology, competitiveness, and MNCs. The paradigm suggests that existence and growth of firms and location of their production will depend, first, on their ability to create and maintain advantages over competitors in the generation of income; second, on the ability of the firms to make the best use of those advantages by extending their own activities or by selling their right of use to other firms; and third, on their ability to choose the most effective locations to create advantages and to produce value-added activity.

Chapters: International Business in a Changing World Environment; Market Power of the Firm and International Transfer of Technology; Multinational Enterprises, Technology and National Competitiveness; Inward Direct Investment from the U.S. and Europe's Technological Competitiveness; The Changing Role of MNEs in the Creation and Diffusion of Technology; International Direct Investment in Innovation: The Pharmaceutical Industry; The Consequences of the International Transfer of Technology by MNEs: A Home Country Perspective; Multinational Enterprises, Industrial Restructuring and Competitiveness: A UK Perspective; The UK's International Direct Investment Position in the Mid-1980s; The Anglo-American Connection and Global Competition; and Multinational Enterprises and the Organization of Economic Interdependence.

293. Enos, John. 1982. "The Choice of Technique vs. the Choice of Beneficiary: What the Third World Chooses." Pp. 69-82 in *The Economics of New Technology in Developing Countries*, edited by Frances Stewart and Jeffrey James. Boulder: Westview Press.

In developing countries, the main choice is among beneficiaries rather than among techniques. Petroleum industry in South Korea as example. Production function approach focuses attention on the wrong variables (choice of technique as the main issue and factor prices as the selection mechanism). Terms of technology transfer contract were the most important in Korean government's choice of technology.

294. Ernst, Dieter. 1988. "Innovation, Technology, and the World Economy: Research and Policy Implications from a European Perspective." Pp. 55-67 in *Science, Technology, and Development*, edited by Atul Wad. Boulder: Westview Press.

Focuses on the role of microelectronics in restructuring international economic relations. Emphasizes four aspects of the impact of microelectronics: (1) instrumentalization of technology for global competition, (2) the "widely exaggerated" expectations of technology, (3) new possibilities for rationalizing economic activities, opened up by microelectronics, and (4) increasing concentration and inequality in key segments of the market. Regarding the first aspect, technology choice and application are subordinated to the demands of global competition. Regarding the second, competitive strategies, such as cheap finance and pricing policies, play a more important role in global competition than technology leadership; however, the informational abilities of microelectronics can make possible rationalization and radical restructuring of economic sectors and allocation patterns. The crucial issue is the increasingly unequal access to information and innovative capacities. Concludes that policies of "high-tech neomercantilism" are leading to a dangerous impasse by undercutting demand, and that refocusing growth on internal demand (demand within nations) could open up new application areas for microelectronics.

295. Fairchild, Loretta G. 1977. "Performance and Technology of United States and National Firms in Mexico." *Journal of Development Studies* 14: 14-34.

A study pairing locally owned and U.S.-owned firms in Mexico. Finds that Mexican manufacturers were competing successfully during the 1966-73 period through "innovative activity" and domestic consulting rather than through formal technology importation. Firm success is measured using profits, sales and assets indicators.

296. Fan, S.G. 1991. "Effects of Technological Change and Institutional Reform on Production Growth in Chinese Agriculture." *American Journal of Agricultural Economics* 73: 266-275.

Identifies the three main determinants of recent rapid growth in Chinese agriculture: increase in inputs, technological change, and institutional reform. Uses an accounting approach to separate the relative contribution of the three factors. Finds that the traditional inputs of land, labor, and manurial fertilizer are still important to China's agriculture, but their relative importance is decreasing rapidly. Modern inputs, such as chemical fertilizer and machinery inputs, played only a very small role in 1965, but by 1985 these had become as important as the traditional inputs.

The most important institutional reform was the household production responsibility system, but regional differences in performance were large and land-scarce regions have gained more from reform. Total input growth was found to explain 57.7% of total production growth, with chemical fertilizer the most important source of growth, increased machinery input the second most important, and traditional inputs less important than these but still significant.

Technological change and efficiency improvement, in the residual, accounted for 42.3% of total production growth. Institutional change has had greater effects on productivity and production growth than technological change.

297. Fransman, Martin and Kenneth King (eds.). 1984. *Technological Capability in the Third World.* New York: St. Martin's Press.

Important volume of essays, most of high quality. Combines a hefty general discussion of Third World technological capabilities, the role of international economics, and work organization with a number of case studies on India, Africa, South Korea, Hong Kong, Brazil, Kenya, and Tanzania.

298. Freeman, Christopher, J. Clark, and L. Soete. 1982. *Unemployment and Technical Innovation: A Study of Long Waves and Economic Development.* London: Frances Pinter.

Examines statistical evidence of changes in the international economy over a long period in the context of major technical and organizational innovations. Following Schumpeter, the authors argue that major inventions and innovations have been unevenly distributed in time and in effects on sectors of industry and services. These "bunched" innovations lead to rapid rises and declines of industries and technologies. Discusses the relation between innovation clusters and long cycles of growth and stagnation and stresses the importance of the diffusion process in the relation between the clustering of innovations and the growth of new industries and the economy. Illustrates the argument with two examples of new technology systems--synthetic materials and electronics.

299. Galal, Essam E. 1983. "National Production of Drugs: Egypt." *World Development* 11: 237-41.

Outlines evolution of the industry in Egypt, and investigates the impact of socio-economic factors on this development. Concludes that, for economic reasons, Egypt has few incentives for R&D and needs to set clear objective goals for this industry, with efficiency and quality competitiveness as high priorities.

300. Gang, Ira N. and Shubhashis Gangopadhyay. 1987.

"Employment, Output and the Choice of Techniques: The Trade-Off Revisited." ***Journal of Development Economics*** **25: 321-7.**

Looks at the question of choice of techniques using a model of an economy with two sectors--urban manufacturing and rural agriculture. More labor-intensive techniques can result in greater unemployment. This is based on four propositions: (1) an increase in urban labor intensity increases the rate of urban employment and the agricultural wage, (2) total unemployment in the economy will increase with the adoption of a more labor-intensive urban technique of production only under certain conditions, (3) for any given size of the urban sector, the higher the elasticity of labor demand in agriculture, the greater the amount of total unemployment in the economy, and (4) any increase in the labor-capital ratio of the agricultural sector increases total unemployment.

301. Ghandour, Marwan and Jurgen Muller. 1977. "A New Approach to Technological Dualism." ***Economic Development and Cultural Change*** **25: 629-38.**

Challenges the two dominant views on technological dualism, identified as the "surplus labor/agricultural surplus absorption" approach and the "factor-markets imperfection" approach. In the alternative interpretation technological dualism is defined as the existence of two production functions for the same commodity facing the respective sectoral entrepreneurs.

302. Goldar, Bishwanath. 1989. "Determinants of India's Export Performance in Engineering Products, 1960-79." ***The Developing Economies*** **27: 3-18.**

Examines the effect of productivity increases on India's export performance in engineering products. Finds that world demand, cumulative output, exchange rate, and total factor productivity are important determinants of export performance. Domestic demand also might affect export performance adversely, but results do not support the hypothesis that higher productivity leads to better export performance.

303. Grabowski, Richard. 1979. "The Implications of an Induced Innovation Model." ***Economic Development and Cultural Change*** **27: 723-34.**

See also volume 30(1) comment and reply. Explores some of the implications of the model of induced innovation proposed by Yujiro Hayami and Vernon Ruttan in their 1971 book. Finds disproportionate advantages to large landowners.

304. Gruebler, Arnulf and Helga Nowotny. 1990. "Towards the Fifth Kondratiev Upswing: Elements of an Emerging New Growth Phase and Possible Development Trajectories." ***International Journal of Technology Management*** **5: 431-471.**

Since the 1970s the world is on the verge of a transition towards a new phase of socio-economic development. A set of new technologies, institutions, and forms of organization will be diffused, leading to renewed growth and prosperity. This may be an opportunity to reduce the disparities between rich and poor countries. The four earlier "waves" are summarized, before introducing new developments in energy, manufacturing, transport and communication. Final remarks on the role of science and technology in the development process.

305. Havrylyshyn, Oli. 1985. "The Direction of Developing Country Trade: Empirical Evidence of Differences Between South-South and South-North Trade." ***Journal of Development Economics*** **19: 255-281.**

South-South trade is economically less efficient because it requires more manpower.

306. Havrylyshyn, Oli and Engin Civan. 1985. "Intra-Industry Trade Among Developing Countries." ***Journal of Development Economics*** **18: 253-71.**

Analyzes intra-industry trade among developing countries and compares this to the intra-industry trade in developed countries. Analyzes IIT among developing countries and among NICs at the product level. Finds that trade of individual NICs with other NICs is below 10% of total exports, which is much lower than developed countries. The level of IIT among NICs is lower than for NICs with the rest of the world. IIT is high in capital-intensive goods and investment products.

307. Huff, W.G. 1989. "Entrepreneurship and Economic Development in Less Developed Countries." ***Business History*** **31:**

86-97.

Essay-length review of two books on developing country entrepreneurship--*Entrepreneurship in the Third World: Risk and Uncertainty in Industry in Pakistan*, by Zafar Altaf, and *Small Firms in Singapore*, by Chew Soon Beng. Interest in LDC entrepreneurship is reviving, but in both Pakistan and Singapore is heavily dependent on the government.

308. James, Jeffrey. 1988. "The Output and Employment Impact of Microelectronics in the Third World: Some Conceptual Issues." Pp. 191-209 in *Science, Technology and Development*, edited by Atul Wad. Boulder: Westview Press.

Suggests a framework to address questions of output and employment effects associated with the use of microelectronics-based innovations in the industrial sector of developing countries. This framework has two components: partial equilibrium analysis and general equilibrium analysis. Partial equilibrium analysis focuses on determining equilibrium price and quantity when the market is viewed as self-contained. General equilibrium analysis considers the effects of technical change in one sector on output and factor markets in other sectors. From the point of view of partial equilibrium analysis, the role of foreign trade (exports) is seen as especially important. As yet little is known about the general equilibrium mechanisms of developing countries, and more empirical work is required in this area.

309. Jha, R., M.N. Murty, S. Paul, B.S. Sahni. 1991. "Cost Structure of India's Iron and Steel Industry: Allocative Efficiency, Economies of Scale, and Biased Technical Progress." *Resources Policy* 17: 22-30.

Analyzes structure of steel costs in India, with aggregate data from 1960 to 1983. Finds that costs have been minimized by factor combinations. By estimating a generalized cost function the authors find that (1) returns to scale have been increasing, so that the industry can significantly reduce cost by increasing input, (2) possibilities for substitution exist among the factors of production; in particular labor and capital are good substitutes, as well as labor and a combination of energy and materials, (3) technical progress has been biased toward the use of labor and energy/materials rather than capital. This last finding means that workers have benefitted from income distribution, relative to suppliers of capital and energy/materials.

310. Juma, Calestous. 1985. "Market Restructuring and Technology Acquisition: Power Alcohol in Kenya and Zimbabwe." ***Development and Change*** **16: 39-60.**

Examines the acquisition of technology for producing power alcohol in these two non-oil producing countries. Suggests that the rapid rate of technological change under current international economic conditions forces countries to adopt dynamic and active strategies for acquiring technology and that theories are needed that can offer flexible and anticipatory analytical frameworks.

311. Justman, Moshe and Morris Teubal. 1990. "The Structuralist Perspective to Economic Growth and Development: Conceptual Foundations and Policy Implications." Pp. 43-69 in ***Science and Technology: Lessons for Development Policy,*** **edited by Robert E. Evenson and Gustav Ranis. Boulder & San Francisco: Westview Press.**

Growth is not an incremental process of capital accumulation (as in the neoclassical view), but involves discrete structural change. Analyzes the structuralist perspective in terms of (1) attitude toward conditions for structural change, (2) comparative advantage, (3) tangible and intangible resource accumulation, and (4) government policies. Cites the initiation of activities of a new industrial sector, such as high technology industry, as the classic case of structural change, but suggests that the notion of structural change be extended to include the introduction and diffusion of new technologies into an economy, restructuring of existing industries, and regional aspects such as decentralization. Discusses a formal model of structural change. Offers a framework for industry and technology policies for the eighties and nineties.

312. Justman, Moshe and Morris Teubal. 1991. "A Structuralist Perspective on the Role of Technology in Economic Growth and Development." ***World Development*** **19: 1167-84.**

Presents a structuralist, rather than orthodox or neoclassical, perspective on economic growth and development. The structuralist view sees structural changes as the cause of growth, rather than capital accumulation and rising per capita incomes. Shows that market failures could be a pervasive phenomenon in a dynamic setting of growth with structural change, rather than an exception as implied by the orthodox neoclassical view.

313. Kibria, M.G. and C.A. Tisdell. 1983. "An Analysis of Technological Change in Jute Weaving in Bangladesh, 1954/55 to 1979/80." ***Developing Economies*** **21: 149-59.**

Discusses the impact of increased capital intensity on production abilities and functions of the industry.

314. Koizumi, Tetsumori and Kenneth J. Kopecky. 1980. "Foreign Direct Investment, Technology Transfer, and Domestic Employment Effects." ***Journal of International Economics*** **10: 1-20.**

Investigates the relationship between foreign direct investment and domestic employment opportunities within the context of multinational firms involved in transferring technology from domestic to foreign operations. The transfer of technology takes the form of a managerial input, which reflects the benefits of learning-by-doing acquired through cumulative gross domestic investment. The model predicts a positive employment effect over the long run because of a favorable interaction between the firm's domestic and foreign operation as a result of the transfer of managerial expertise.

315. Krugman, Paul. 1979. "Model of Innovation, Technology Transfer, and the World Distribution of Income." ***Journal of Political Economy*** **87: 253-66.**

How technological change relates to international trade. General-equilibrium model of product cycle trade, based on products developed in one country and produced, after a lag, in a second country. The first country exports new products and imports old products.

316. Lalkaka, R. 1988. "Is the United States Losing Technological Influence in the Developing Countries?" ***Annals of the American Academy of Political and Social Science*** **500: 33-50.**

Finds that Third World countries generally are more receptive to American technology and investment than in the past, a change important for the U.S. since it is losing influence over the world's high-technology markets and needs developing countries for markets and raw materials.

317. Lall, Sanjaya. 1982. *Developing Countries as Exporters of Technology: A First Look at the Indian Experience.* London: Macmillan.

Considers an important and generally neglected topic, the extent to which developing countries can become internationally competitive exporters of technology. Uses India as a case study because its success is not as phenomenal as that of the other NICs, because it employs a more "inward-looking" strategy than most, and because it exports more varied technologies. Includes a discussion of the concept of "technology exports" and review of evidence from India, including some limited comparisons with other countries. Attempts to account for comparative advantage.

318. Lall, Sanjaya and Rajiv Kumar. 1981. "Firm-Level Export Performance in an Inward-Looking Economy: The Indian Engineering Industry." *World Development* 9: 453-64.

Examines the relationship between R&D and export performance for the 100 largest engineering enterprises in India. Finds that exporting is negatively correlated with profitability and technological activity, although some individual firms show a relation between R&D and exporting over time. Finds some signs of change, with indications that exporting may be becoming more profitable for Indian industries. The appendix contains firm-level data from 1976 to 1978, with figures on sales, exports, and profits before tax.

319. Lall, Sanjaya and Sharif Mohammad. 1983. "Technological Effort and Disembodied Technology Exports: An Econometric Analysis of Inter-Industry Variations in India." *World Development* 11: 527-35.

Tests determinants of Indian technological capability by studying "disembodied" exports arranged by the largest private sector companies in India. Concludes that: (1) Formal R&D is a significant positive factor in technological development as embodied in technological exports, (2) Skill and scale requirements appear to promote, not repress, technological development, and (3) The presence of foreign ownership appears not to inhibit technological development or technological exports.

320. Levy, V. 1981. "Total Factor Productivity, Non-Neutral Technical Change and Economics Growth: A Parametric Study of a Developing Economy." ***Journal of Development Economics*** **8: 93-?**

Analyzes the manufacturing sector of Iraq and attempts to measure the contributions of technological progress and primary factors of production to economic growth.

321. Lin, J.Y. 1991. "Public Research Resource Allocation in Chinese Agriculture: A Test of Induced Technological Innovation Hypotheses." ***Economic Development and Cultural Change*** **40: 55-74.**

Attempts to test four hypotheses of induced technological innovation: (1) a decentralized public research system will allocate more resources to developing technology that saves scarce factors even in a controlled economy, (2) in this institutional environment, public research institutes will allocate more resources to crops with larger market demand, (3) varieties of a given crop in a region increases as the resources allocated to variety-improvement research increase, and (4) the availability of new varieties of a crop developed by public research institutes in a region is positively correlated with market demand of that crop in the region. Uses empirical data collected from China's public agricultural research institutes. Evidence indicates that the decentralized system of agricultural research in China has allocated resources in a way consistent with predictions of both the factor-scarcity-induced and the market-demand-induced technological innovation hypotheses.

322. Long, Frank. 1978. "Basic Needs Strategy for Development of Technology in Low Income Countries." ***American Journal of Economics and Sociology*** **37: 261-70.**

Advances the case for an "operationally-oriented basic needs concept" for technology in low income countries and outlines a planning model. Eight equations are developed to show basic needs technology interacting with a structure of production, consumption, and accumulation.

323. Lower, Milton D. 1987. "The Concept of Technology within the Institutionalist Perspective." ***Journal of Economic Issues*** **21: 1147-1176.**

Not specifically about LDCs but applicable. Explanation of

Veblen/Ayres/Dewey paradigm (technology/ceremony dichotomy) organized in terms of technology and cultural evolution, technological progress, and technology and instrumental behavior. Ayres' principle of tool combination.

324. Magee, Stephen P. 1981. "The Appropriability Theory of the Multinational Corporation." ***Annals of the American Academy of Political and Social Sciences*** **458: 123-35.**

The appropriability theory of MNCs emphasizes conflict between innovators and emulators of new technology -- probability of appropriation is substantial when innovator profits are high relative to the innovation's value to society. The most important consideration facing innovators is the possible loss of technology to copiers and rivals. It is more efficient to transfer high tech inside firms than through the market. Mechanisms that evolve to prevent loss of high tech can explain much MNC behavior. MNCs cannot be counted on to create the technology most useful for developing countries.

Appropriability theory emerges from Vernon's product cycle--production occurs in LDCs so late in the product cycle that discounting gives the importance of cheap unskilled LDC labor a small weight to the multinational. Appropriability is the reason why firms that develop new products become large: they expand to internalize the public-goods aspect of new information. Countries cutting the price they pay for existing technology gain in the short run, but in the long run private competition will fail to provide innovations and disseminate them if the legal system and national policy do not balance the economic rights of innovators and emulators.

325. Manrique, Gabriele. 1987. "Intra-Industry Trade Between Developed and Developing Countries: The United States and the NICs." ***Journal of Developing Areas*** **21: 481-94.**

Presents calculations of the share of intra-industry trade in the manufactured goods trade of the United States and the NICs for the years 1967, 1972, 1977, and 1982. Uses cross-section regression analysis of the manufacturing industries included in the 1982 U.S. trade data to test for the determinants of the level of intra-industry trade. Principle conclusions: Intra-industry trade was present even when the NICs were still LDCs (late 70s). Intra-industry trade levels tend to be higher in industries characterized by greater product differentiation and less in concentrated industries. Tariff barriers affected level of trade in some countries. Intra-industry trade tends to be lower in high-wage industries.

326. Marton, K. and R.K. Singh. 1991. "Technology Crisis for Third-World Countries." *World Economy* 14: 199-214.

The authors see flows of industrial technology to developing countries as stagnant and even declining in the latter half of the 1980s. Two important factors in this slowdown have been national policies in developing countries toward foreign technology acquisition and the growing gap in electronics application and usage. Major policy changes, including large capital investments in communications and production equipment and massive development of human resources, will be necessary.

327. Matthews, R. 1987. "The Development of a Local Machinery Industry in Kenya." *Journal of Modern African Studies* 25: 67-93.

Examines the machinery industry in Kenya. This piece is largely descriptive, but contains some policy proposals. Data from the Statistical Abstracts of the Ministry of Economic Planning and Development in Nairobi.

328. Mead, D.C. 1991. "Small Enterprises and Development." *Economic Development and Cultural Change* 39: 409-19.

Critical review essay of *Small Manufacturing Enterprises: A Comparative Analysis of India and Other Economies*, by Ian M.D. Little, Dipak Mazumdar, and John M. Page.

329. Nelson, R.R. and V.D. Norman. 1977. "Technological Change and Factor Mix over the Product Cycle: A Model of Dynamic Comparative Advantage." *Journal of Development Economics* 4: 3-?

Presents a simple dynamic model for a given set of factor prices, showing how the optimizing mix of factor inputs changes over a product cycle, and how, as a result, comparative advantage in international trade shifts.

330. Pack, Howard. 1984. "Productivity and Technical Choice: Applications to the Textile Industry." *Journal of Development Economics* 16: 153-76.

Analyzes past choices of technology and present levels of productivity in the cotton textiles industry (i.e., spinning and weaving) in the Philippines. Uses engineering and economic information to assess costs of alternative technologies, estimate levels of productivity, and analyze sources of productivity shortfalls. Ascribes high production costs to inappropriate technological choices and to low productivity in use of technologies. Does not find low labor skills to be a major problem, but attributes low productivity to lack of sufficient firm-level specialization among product varieties and deficient firm-level technological capabilities.

331. Pack, Howard. 1990. "Industrial Efficiency and Technology Choice." Pp. 209-231 in *Science and Technology: Lessons for Development Policy*, edited by Robert E. Evenson and Gustav Ranis. Boulder: Westview Press.

Analyzes the Philippine and Kenyan textile sector in order to look at the relation of technology choice to industrial efficiency. Arrives at three broad conclusions regarding choice of technology: (1) both radically new and very old designs in equipment are not competitive in the economic environments of these two countries; new conventional equipment offers a cost advantage, (2) used conventional equipment is more advantageous than new conventional equipment at current productivity levels, due to lower price and greater productivity relative to potential, (3) after machines have been chosen, the correct choice of the amount of labor they require may result in a large saving. Suggests three types of policies to improve productivity: (1) liberalization of the trade regime, (2) efforts to rationalize the fragmented production structure, and (3) technical aid to firms to help with engineering and organizational problems.

332. Pack, Howard and Larry E. Westphal. 1986. "Industrial Strategy and Technological Change: Theory versus Reality." *Journal of Development Economics* 22: 87-128.

Industrialization should be thought of as managing technical change, not in terms of invention but in terms of adapting foreign technology. Research demonstrates that market forces alone are not responsible for the success of the East Asian NICs, using Korea as the main case. Trade policies are secondary to technological ones. Criticizes neoclassical arguments for a neutral policy regime, contrasting them with newer views. Firm-level case-study research is reviewed. Industrial products and elements of technology are imperfectly tradable or inherently nontradable.

333. Pickett, James (ed.). 1977. *World Development* 5: 773-882.

Special issue on choice of technology in developing countries. Articles on leather manufacturing and iron foundry, maize milling, brewing, footwear, bolt and nut manufacture, fertilizer and African textiles.

334. Prendergast, R. 1990. "Causes of Multiproduct Production: The Case of the Engineering Industries in Developing Countries." *World Development* 18: 361-70.

Uses a variety of concepts developed in recent literature on multiproduct industries to explain why vertical and horizontal specialization in developing country capital goods industries has not occurred. Maintains that firms generally diversify their product mix because they are seeking to take advantage of idle machine capacity. When they do this, however, they do not take into account the organizational problems caused by diversification. As market size grows, greater specialization will occur as economies of scope resulting from idle machine capacity disappear. The incentive for diversification will continue if domestic prices for a firm's products remain higher than world prices.

335. Roe, Terry and Matthew Shane. 1979. "Export Performance, Marketing Services, and the Technological Characteristics of the Malaysian Industrial Sector." *Journal of Developing Areas* 13: 175-89.

Provides industrial-sector data from Malaysia to show that employment and production in labor-intensive subsectors can be stimulated by lowering an industry's factor (input) costs and/or increasing the market prices of its products. Argues that government policies to increase marketing efficiency will be more effective than those affecting relative factor prices.

336. Roemer, M. 1979. "Resource-Based Industrialization in the Developing Countries: A Survey." *Journal of Development Economics* 6: 163-202.

Attempts to shed light on the potential contribution of resource processing to efficient growth, employment creation, greater equity, and economic independence. Suggests that because resource-based industries are not impressive contributors to direct or indirect employment creation, they are likely to perpetuate the pattern of dualism and inequality present in typical resource-rich countries.

337. Rosenberg, Nathan. 1990. "Science and Technology Policy for the Asian NICs: Lessons from Economic History." Pp. 135-155 in *Science and Technology: Lessons for Development Policy*, edited by Robert E. Evenson and Gustav Ranis. Boulder & San Francisco: Westview Press.

Looks at American industrial history to find lessons in S&T policy for the NICs. Concerned with how science and technology have exercised their influence, what factors have determined their effectiveness, and what have been the connections between science and technology. Finds four significant features in the American experience: (1) a roughly simultaneous growth in research activities in private industry, academia, and the public sector, suggesting a common underlying force, (2) compared with Europe, a much larger fraction of business-supported research in the U.S. was conducted within the firm, rather than by industry-wide associations, (3) well-defined limits to the extent to which a firm can overcome deficiencies by relying on services available through the market, and (4) successful institutions have approached problems of creating interactions among specialists in different ways. Suggests that the critical question for LDCs is how to link applied science with improved technologies and improvements in the performance of economies. Cites India as an example of a country that has received low payoffs from a well-developed S&T infrastructure. Looks briefly at the relevance of the Japanese historical experience for the NICs.

338. Ruttan, Vernon W. and Yujiro Hayami. 1984. "Toward a Theory of Induced Institutional Innovation." *Journal of Development Studies* 20: 203-23.

Marshals evidence to suggest institutional innovation is induced through changes in relative resource endowments that lead to changes in relative factor prices and cost-reducing technologies, using examples from agricultural history. Accuses neo-Marxist critics of the Green

Revolution of overly stressing institutional change effected by technology and neglecting changes in resource endowments brought about through population increase. The authors also argue the significance of "cultural endowments," which they say social sciences conceal under the rubric "taste" or "ideology," but call attempts to measure such endowments unsatisfactory.

339. Sagasti, Francisco. 1988. "Market Structure and Technological Behavior in Developing Countries." Pp. 149-168 in *Science, Technology and Development*, edited by Atul Wad. Boulder: Westview Press.

Looks at the factors that influence technology choice and technological behavior in industry. Analyzes market structure, type of technology, and characteristics of the leading firms in a branch. Examines interactions among these factors and the balance between profit and net nonpecuniary benefits, and considers implications of these interactions for patterns of competition and strategies of individual firms.

340. Samuels, B.C. 1990. *Managing Risk in Developing Countries: National Demands and Multinational Response*. Princeton, NJ: Princeton University Press.

Examines country-risk management through case studies of conflict between MNC and host-country demands in the Brazilian and Mexican automotive industries in the early 1980s. Based mostly on interviews with executives in MNC subsidiaries. In Brazil, LDC groups pressured MNC subsidiaries for changes in local labor policies while in Mexico the host government established regulations that affected the investment strategy of the parent companies. Subsidiaries are viewed as "floating" between the parent company and the host company, subject to competing claims. Evidence is partially supportive of both dependency and pro-MNC theories.

341. Scott, A.J. 1987. "The Semiconductor Industry in South East Asia: Organization, Location and the International Division of Labor." *Regional Studies* 21: 143-60.

Deals with the functional and geographical organization of the semiconductor industry in Southeast Asia. Includes a broad description of the growth of U.S.-owned assembly plants, locally owned diffusion facilities, and locally owned subcontract assembly houses. The internal functional characteristics of assembly plants are examined. An international division of labor is emerging within Southeast Asia. The growth of semiconductor production complexes is analyzed. The theory of the new international division of labor is discussed.

342. Sharma, S.C. 1991. "Technological Change and Elasticities of Substitution in Korean Agriculture." ***Journal of Development Economics*** **35: 147-72.**

Uses translog cost function to investigate two issues concerning the Korean economy during the prewar (1918-1938) and postwar (1949-1979) periods: (1) analyzes characteristics of the technology used and the process of technical innovation; (2) attempts to determine the partial elasticities of substitution for pairwise comparisons of land, labor, fixed capital, and working capital. Finds that technological innovations in Korean agriculture were land- and labor-saving innovations, fixed-capital neutral innovations, and innovations using working capital. Finds that working capital was a strong substitute for land in the prewar period and a weak substitute in the postwar period. Fixed capital is found to have been a weak substitute for land and labor in the prewar period and a weak complement to them after the war. Working capital and fixed capital are found to have been weak substitutes before the war and strong complements after the war.

343. Skorov, G.E. (ed.). 1978. ***Science, Technology and Economic Growth in Developing Countries.*** **Oxford: Pergamon.**

A collection of translated essays from the USSR Academy of Sciences illuminating the role of science and technology in social and economic development in newly independent countries. Essays look at barriers to progress, measures to combat technical and economic backwardness, and cooperation with socialist countries. A "socialist" perspective and projections for progress in the Third World.

344. Soete, L. 1981. "Technological Dependency: A Critical View." Pp. in ***Dependency Theory: A Critical Reassessment*****, edited by D. Seers. London: Frances Pinter.**

A critique of the dependency approach to technology in developing countries. Argues that this approach ignores the dynamic aspects of technology. Discusses the economics of dependency and suggests that the "terms of trade" concept (i.e. deteriorating terms of trade for developing countries with regard to developed countries) is probably the most useful dependency concept. Develops a classification scheme for elements of dependency in terms of industrial organization. In this scheme products are classified according to level of scarcity/degree of substitutability, number of sellers and type of dependence created. Brief statement of the technological dependency argument and a systematic examination of the shortcomings of that argument. Maintains that continuous and massive technological dependency is actually the best way for developing countries to catch up with the North.

345. Solow, Robert. 1957. "Technical Change and the Aggregate Production Function." *Review of Economic Statistics* 39: 312.

Important neoclassical growth-accounting model, indicating that technical change has to be considered a major determinant of economic growth in the U.S. Estimates that the contribution of S&T to the 20th-century U.S. GNP ranges from 40% to 60%.

346. Stewart, Frances and J. James (eds.). 1982. *The Economics of New Technology in Developing Countries*. London: Frances Pinter.

Excellent collection of essays (many of which have been published in other forms) offering "new perspectives" on technological change, in contrast to the neoclassical question of the labor- or capital-intensity of production technique. These take the point of view that Third World countries should develop their own capacity to create technology, owing to the fact that imported technology tends not to be well adapted to Third World needs, to the inappropriateness of imported technology to local circumstances, and to the need to reduce technical dependence on developed countries. Chapters by prominent scholars on appropriate technology, infant industries, technology exports in the Third World, the Indian tractor industry, agricultural research, welfare effects of new products, and product standards in developing countries.

347. Tho, Tran Van. 1988. "Foreign Capital and Technology in the Process of Catching Up by the Developing Countries: The Experience of the Synthetic Fiber Industry in the Republic of Korea." *Developing Economies* 26: 386-402.

Maintains that the major factor in changes to North-South trade has been the attempt by developing countries to catch up with developed countries in manufacturing. This paper looks at how developing countries have attempted to catch up by substituting local resources for foreign capital, technology, and other managerial resources. Case study of the synthetic fiber industry in the Republic of Korea. Korea first had to depend on Japanese technology and capital but later was able to substitute its own. Two implications are drawn from the Korean experience: (1) a developing country can exploit the advantages of being a developing country by using as much foreign capital, technology, and managerial knowledge as possible and (2) the developing country should try to absorb the imported managerial resources efficiently and gradually replace them with its own.

348. Weeks, John and Elizabeth Dore. 1979. "International Exchange and the Causes of Backwardness." ***Latin American Perspectives*** **6: 62-87.**

Argues from a Marxist standpoint that exploitation is a relationship between classes, not between countries. Suggests that inequality among countries results from processes of exploitation within "backward" countries, processes that are spread through imperialism. Denies that the negative effect of international commerce in Third World countries is a result of the terms of trade.

349. Weiss, Charles, Jr. 1990. "Scientific and Technological Constraints to Economic Growth and Equity." Pp. 17-41 in ***Science and Technology: Lessons for Development Policy*****, edited by Robert E. Evenson and Gustav Ranis. Boulder & San Francisco: Westview Press.**

Nice overview of the role of science and technology in achieving the development objectives of growth, employment, and equity. Distinguishes six critical aspects of S&T development that may constrain the contribution of S&T to economic growth and equity: (1) general economic development, (2) capacity of the productive sector to manage technology, (3) the state of market-oriented technology policy, (4) availability of money, (5) availability of people at all levels of training, and (6) the scientific and technological infrastructure.

Presents a model of market-oriented scientific and technological development, in the form of a series of stages. These consist of two broad stages: the emergence of islands of modernization and the struggle for mobilization and mastery. Gives salient features of each stage. Each is divided into three substages: (1a) traditional technology, (1b) first emergence, (1c) islands of modernization, (2a) mastery of conventional technology, (2b) transition to newly industrializing country, and (2c) threshold of technological competence. Gives examples of countries at each substage. Discusses nonmarket paths to S&T development and external factors that may affect the contribution of S&T. Examines investment in S&T and the role of development assistance.second page text starts here

Chapter 5 -- Innovation

350. Agarwal, Bina. 1983. "Diffusion of Rural Innovations: Some Analytical Issues and the Case of Wood-burning Stoves." ***World Development*** **11: 359-76.**

Shows how involving the users in development is an effective strategy, and how top-down diffusion is not. Maintains that the actual nature of the innovation determines what diffusion strategy should be used.

351. Ahmed, Iftikhar. 1985. "Rural Women and Technical Change: Theory, Empirical Analysis and Operational Projects." ***Labour and Society*** **10: 289-306.**

Discusses the mechanisms and causes of female labor displacement, with a focus on Africa. Technological change generally leads to concentration of women in domestic and non-market roles.

352. Ahmed, I. and V.W. Ruttan. 1988. ***Generation and Diffusion of Agricultural Innovations: The Role of Institutional Factors.*** **Brookfield, VT: Gower Publishing Co.**

A collection of essays examining agricultural technology diffusion, with specific case studies on India, Bangladesh, Kenya and Vietnam. Emphasizes the non-transferability of agricultural innovations. Essays address four broad issues: (1) testing of innovation theories, (2) the effect of farm-sized distribution, (3) the effect of decentralized research on the supply of new technology to farmers, and (4) the biases in agricultural research and extension programs.

353. Aina, Lenrie O. 1986. "Agricultural Information Provision in Nigeria." ***Information Development*** **2: 242-4.**

Looks at why Nigerian farmers have received inadequate information on the results of agricultural research. Suggests that extension officers need to strengthen their information capabilities through the provision of information by librarians. (Author is a university lecturer in library sciences.)

354. Bagchi, A.K. 1990. "Some Fundamental Issues in the Analysis of Technological Change in Developing Economies." ***Social Science Information*** **29: 398.**

Maintains that analysis of technological change in advanced capitalist countries has taken the form of a trichotomous division: "Invention" (basic ideas about a new product or new process are formulated), "Innovation" (invention is put to commercial use), and "Diffusion" (invention is adopted by persons or firms other than the original innovator or inventor). This division is not relevant to LDCs, where technical change is mainly a process of importing technologies, processes, and products.

Necessary components of a study of technical change in LDCs must include: (1) study of the process of transfer from DCs to LDCs, (2) process of absorption of transferred technologies, (3) course of diffusion and adaptation of the new technologies to the local environment, (4) mutual interaction of diffusion and adaptive development, and (5) diffusion of old technologies that may have become more economic in the new environment. Presents a typology of developing economies based on 10 characteristics. Discusses organizational and infrastructural factors influencing the impact of technology. Considers the relationship between technical change, structural change, and social change.

355. Barnett, A. 1990. "The Diffusion of Energy Technology in the Rural Areas of Developing Countries: A Synthesis of Recent Experience." ***World Development*** **18: 539-54.**

Looks at how the perspective on technology diffusion has shifted in recent years from social-psychological concerns to concerns about the nature of the technology itself, the political economy of users and producers, and more "evolutionary" views of the process of technical change. The article introduces four related articles in this journal. Concludes by offering a summary of basic principles to guide strategies for more productive use of energy resources in rural areas: (1) invest more in understanding needs of potential users in rural areas, (2) consider a wide range of options for meeting each of these needs in their particular location, (3) make more of an effort to understand how macro policy environment will affect the introduction of technology, (4) develop simple, reliable technology that can satisfy numerous users, needs and locations, and avoid experimental and untried technology unless it is part of an evaluative program that can monitor progress and draw necessary lessons, (5) ensure long-term commitment to meeting the chosen energy end use, (6) harmonize diffusion strategy with local physical, human, and institutional resources, and (7) build a production, delivery, and maintenance system that can monitor progress and adapt to changing circumstances.

356. Bernardo, Francisco P. and Gregorio U. Kilayko. 1990. "Promoting Rural Energy Technology: The Case of Gasifiers in the Philippines." *World Development* 18: 565-574.

Evaluates the diffusion of gasifier technology in the Philippines. This is based on a field survey of 53 farmers' groups using gasifiers on Panay Island in 1985, and it is supplemented with information on the entire country that became available in 1987. Concludes that problems in the diffusion of gasifiers were institutional rather than technical.

357. Bhattia, Ramesh. 1990. "Diffusion of Renewable Energy Technologies in Developing Countries: A Case Study of Biogas Engines in India." *World Development* 18: 575-590.

Studies the diffusion of biogas engines in India in terms of the characteristics of the technology, the characteristics of users, the macro environment, and the role of government and other agencies. Although biogas engines are known to Indian farmers, this technology has not been widely adopted. A major reason for this has been the large difference in prices between biogas and diesel engines. The author suggests that investment in biogas engines should be encouraged by raising prices of diesel oil and electricity or by giving investment supports and subsidies to biogas equal to those for diesel oil and electricity.

358. Biggs, Stephen D. and Edward J. Clay. 1981. "Sources of Innovation in Agricultural Technology." *World Development* 9: 321-36.

Looks at formal and informal means of R&D and gauges their efficiency, focusing on biological and environmental characteristics of agriculture that shape the process of innovation. Case studies of farmers engaged in the continuous process of innovation, using dwarf wheat as an example of formal process. Five policy issues reviewed: genetic vulnerability, the development of environmentally specific or widely adapted technologies, optimal location of experiment stations, importance of an on-farm research component in formal R&D systems, and need to strengthen and complement informal systems.

359. Blumenthal, Tuvia. 1979. "A Note on the Relationship between Domestic Research and Development and Imports of Technology." *Economic Development and Cultural Change* 27: 303-6.

Examines the relationship between domestic and imported technology and attempts to find out whether substitution or complementarity is dominant. Uses cross-sectional data of industries in several countries. Do countries which are large importers of foreign know-how also engage in massive R&D programs? On the whole, the relationship is found to be one of complementarity.

360. Braverman, Avishay and Joseph E. Stiglitz. 1986. "Landlords, Tenants and Technological Innovations." *Journal of Development Economics* 23: 313-32.

Landlords suppress innovations that would lighten the debt load of tenants, but sometimes adopt innovations that increase it. General discussion of institutional structure and its relationship to innovation adoption.

361. Cantwell, John. 1989. *Technological Innovation and Multinational Corporations*. Oxford: Basil Blackwell.

Maintains that since 1945 patterns of international trade and production have been shaped by competition to accumulate technology. Suggests that the spread of American multinationals stimulated technological innovation in Europe and, in turn, European multinationals expanded in America. Since technological knowledge must be generated by firms producing in international centers of innovation, on-site production is encouraged. Developing countries receive only minor attention.

362. Chudnovsky, Daniel. 1988. "The Diffusion and Production of Numerically Controlled Machine Tools with Special Reference to Argentina." *World Development* 16: 723-32.

Shows how Third World firms are adopting NCMTs, a form of automated machine-building technology, despite cheap labor markets. The author describes this new technology, in which numbers form a program of instructions designed for a particular job and instructions can be easily changed as the job changes. NCMTs were developed in the U.S. but have been diffusing due both to declining costs of the technology and expansion of the user market.

Diffusion of NCMTs in developing countries has occurred mainly where capital goods are being produced and where the conventional machine tool industry is already well established. Most NCMTs in developing countries are imported from industrialized nations, although Brazil produces a significant number. In Argentina, the technology began to be adopted in 1979-81 when wages were high and imported capital goods were cheap, but NCMTs continued to be installed even after wages dropped, indicating that their use is linked more to the expansion into production of more complex machinery than to relative costs of capital and labor. However, diffusion may be limited by the technology's requirement of a high aggregate volume of production to amortize costs, unless purchase of NCMTs is subsidized by developed countries. Production of NCMTs in Taiwan, Brazil, and South Korea has been made possible by tariff protection, import restrictions, and other forms of government involvement. The history of Argentina's production of this technology since 1979 is described.

363. Clark, Norman. 1990. "The International Diffusion of Biotechnology: Some Ideas for a Common Research Framework." Pp. 77-90 in *Science Policy Research: Implications and Applications*, edited by D.F. Dealmeida. London: Pinter Publishers Ltd.

A general discussion of methodological problems faced in the analysis of ongoing, rapidly changing policy decisions. Barriers to research include the inability to conduct actual experiments, the long-term nature of research policy allocations, and the "lack of coherent S&T policy" because technologies straddle multiple economic sectors.

364. Colle, R.D. 1989. "Communicating Scientific Knowledge." Pp. 59-83 in *The Transformation of International Agricultural Research and Development*, edited by J. Lin Compton. Boulder & London: Lynne Rienner Publishers.

Looks at the process of transferring scientific information from scientists to farmers and how this can be done quickly enough so that farmers will get the greatest possible benefit. Reviews diffusion studies, examines communication problems that have arisen from new communication media, from new types of clientele for extension agencies, and from scientists' inefficiency in communication. Suggests approaches to improving communication.

365. Compton, J. Lin. 1989. "The Integration of Research and Extension." Pp. 113-136 in *The Transformation of International Agricultural Research and Development*, edited by J. Lin Compton. Boulder & London: Lynne Rienner Publishers.

How international agricultural R&D programs work. Brief overview of past developments, such as the package program, based on the principle that farmers need a combination of techniques and supports, such as credits, markets, irrigation, and inputs; the induced innovation approach, based on the principle that technical changes can be guided by price signals; and a variety of decentralized models. Considers some of the lessons learned from past experience, suggesting that the integration of agricultural research and extension in Third World countries has resulted in both failures and successes. Briefly examines events in research and extension over the past five years, such as the World Bank's "Training and Visit approach."

366. Dahlman, C. and L.E. Westphal. 1982. "Technological Effort in Industrial Development: An Interpretative Survey of Recent Research." Pp. in *The Economics of New Technology in Developing Countries*, edited by Frances Stewart and Jeffrey James. Boulder, CO: Westview.

Technological mastery concept applied to production engineering, project execution, capital goods manufacture, and R&D. Mastery at any stage contributes to mastery at others. Experience is the key to mastery.

367. Daxiong, Qiu, Gu Shuhua, Liange Baofen, and Wang Gehua. 1990. "Diffusion and Innovation in the Chinese Biogas Program." *World Development* 18: 555-563.

Describes the history of biogas developments in China. Contrasts early phases, which emphasized producing numerous plants, and the current phase, which concentrates on constructing high-quality plants. Reports results of microeconomic analysis of 58 biogas plants in Tongliang and compares this with data of previous researchers on 242 plants in Hubei. Finds high rates of return on investments in biogas and short payback periods of one to four years. Examines the impact of China's institutional reforms on biogas projects and finds that the number of plants built each year has fallen as a result of reductions in subsidies, emphasis on quality rather than quantity, and rising rural incomes which have led to a switch from biogas to coal.

368. Desai, Ashok V. 1988. "Indigenous and Foreign Determinants of Technological Change in Indian Industry." Pp. 400-31 in *Technology Transfer By Multinationals*, edited by H.W. Singer, Neelambar Hatti, and Rameshwar Tandon. New Delhi: Ashish Publishing House.

Four types of technological capability are defined: in the purchase of technology, in plant operations, in duplication and expansion, and in innovation. Discusses concentration of Indian industry and evolution of current market structure. Indian buyers are inhibited by the inability to attract technology sellers of their choice. Engineering industry has been built up through import substitution.

369. Downs, George W., Jr. and Lawrence B. Mohr. 1980. "Toward a Theory of Innovation." Pp. 75-100 in *Innovation Research and Public Policy*, edited by John A. Agnew. Syracuse: Syracuse University Press.

Helpful overview not specifically about developing countries. Aims to describe necessary conditions for formulating a theory of innovation and to develop a model that satisfies these conditions. Argues that any model attempting to describe the extent and time of adoption must include characteristics of the organization and the innovation in context. This approach depends on identifying (1) dimensions of the choice situation (benefits, costs, resources, discounting factors), (2) a theoretical and mathematical framework that links these choices. Presents a mathematical model of a theory of innovation.

370. Eastman, Clyde and James Grieshop. 1989. "Technology Development and Diffusion: Potatoes in Peru." Pp. 33-55 in *The Transformation of International Agricultural Research and Development*, edited by J. Lin Compton. Boulder & London: Lynne

Rienner Publishers.

Looks at Peru as an example of foreign adaptation of the U.S. land-grant model. Describes the evolution of technical capability in the Peruvian Ministry of Agriculture and the National Agrarian University over the course of the 20th century. Also examines how the program of the North Carolina Mission and others influenced agricultural institutions, and how they have affected Peruvian agriculture. Interprets the findings and makes suggestions for future attempts to build agricultural institutions.

371. Fairchild, Loretta G. and K. Sosin. 1986. "Evaluating Differences in Technological Activity Between Transnational and Domestic Firms in Latin America." ***Journal of Development Studies*** **22: 697-708.**

A study of Latin American firms from four regions. Domestic firms and MNCs are roughly equal in growth and profitability. The main difference is that domestic firms have a relatively higher level of internal innovation activity; foreign firms rely heavily on innovation sources external to the host nation.

372. Feder, Gershon and Gerald T. O'Mara. 1981. "Farm Size and the Diffusion of Green Revolution Technology." Economic ***Development and Cultural Change*** **30: 59-76.**

Presents a theoretical decision model and its implications for the optimal behavior of farmers. A specific example serves as a basis for numerical situations tracing the pattern of adoption in a hypothetical rural region over time. Simulations demonstrate the income distribution effects of the innovation over time, showing that income distribution is likely to deteriorate in the initial stages, improving only when smaller farmers join the ranks of adopters.

373. Fitzgerald, Deborah. 1986. "Exporting American Agriculture: The Rockefeller Foundation in Mexico, 1943-53." ***Social Studies of Science*** **16: 456-83.**

Shows how a plan to export U.S. agricultural techniques, which was generally considered a success, actually worked only to the advantage of farmers who were already Americanized in their processes.

374. Fransman, Martin. 1984. "Explaining the Success of the Asian NICs: Incentives and Technology." *IDS Bulletin* 15(2): 50-6.

Suggests the process of innovation must be considered when assessing causes for the success of the Asian NICs, a perspective neglected by neoclassical assessments.

375-376. Gamser, Matthew S. 1988. "Innovation, Technical Assistance, and Development: The Importance of Technology Users." *World Development* 16: 711-21.

Shows how Sudanese programs to increase the feedback from technology users to developers created successful initiatives in charcoal production, charcoal stoves and agroforestry. Suggests that technology users play an essential part in the development of products and processes. New approaches to technical assistance are needed to increase interaction between R&D institutions and technology users. Gives examples of interactive R&D in agriculture (such as when the International Potato Center in Peru reorganized its research after consulting peasants on their problems), health (WHO's use of traditional medicine), and irrigation (the need for farmers to play a role in design and implementation of irrigation systems). Describes how interactive technology development can be managed by suggesting a "biological model," involving direct feedback on research problems, as an alternative to the more widely used "hierarchical model." Summary of user interaction in forest energy technology development in Sudan.

377. Girvan, N.P. and G. Marcelle. 1990. "Overcoming Technological Dependency: The Case of Electric Arc (Jamaica) Ltd., a Small Firm in a Small Developing Country." *World Development* 18: 91-108.

A case study of a successful Jamaican business enterprise. The success is partly attributed to good strategies for acquiring and assimilating technology. Instead of using licensing agreements, Electric Arc used relationships with suppliers of raw materials and equipment, as well as in-plant experimentation and learning. Concludes that active strategies such as this are best for successful technology acquisition.

378. Goodell, Grace E. 1984. "Bugs, Bunds, Banks, and Bottlenecks: Organizational Contradictions of the New Rice Technology." *Economic Development and Cultural Change* 33: 23-42.

Maintains that changes in social organization are required for the adoption of new rice technology. Focuses on field-level organizational problems posed by new irrigated rice technologies in South and Southeast Asia. Suggests that we need more active participation from Third World farmers in the research and agricultural policy-making carried out for their benefit, so that anthropologists and others will not have to serve as middlemen between users and scientists/government.

379. Goonatilake, Susantha. 1987. "Inventions and the Developing Countries." ***Impact of Science on Society*** **147: 223-31.**

A highly theoretical piece that identifies three relationships that must be examined to understand the lack of invention in developing economies--that between discovery and invention, between invention and culture, and between invention and socio-economic processes. The discussion especially concerns biotechnology and information technology. Argues that several systems of technology and several modes of production may coexist, and that some may be particularly suited to different modes of thought. The development of artificial intelligence may be particularly suited to Eastern models of mental processing.

380. Gore, A.P. and U.A. Lavaraj. 1987. "Innovation Diffusion in a Heterogeneous Population." ***Technological Forecasting and Social Change*** **32: 163-8.**

Suggests that diffusion is hampered when a population is differentiated along some critical dimension, and offers a model to anticipate the difficulties.

381. Herrera, Amilcar O. 1981. "The Generation of Technologies in Rural Areas." ***World Development*** **9: 21-36.**

Technologies are developed now under a paradigm constructed for developed countries. Proposes a methodology for developing countries that could lead to a Third World paradigm.

382. Hicks, W. Whitney and S.R. Johnson. 1979. "Population Growth and the Adoption of New Technology in Colonial Taiwanese Agriculture." ***Journal of Development Studies*** **15: 289-303.**

Argues that the acceptance of labor-using, high-yield

technologies followed population growth.

383. Hill, Hal. 1985. "Subcontracting, Technological Diffusion, and the Development of Small Enterprise in Philippine Enterprise." *Journal of Developing Areas* 19: 245-62.

Discusses a governmental attempt to construct subcontracting networks for technology diffusion, which experienced limited success.

384. Hurst, Christopher. 1990. "Establishing New Markets for Mature Energy Equipment in Developing Countries: Experience with Windmills, Hydro-powered Mills and Solar Water Heaters." *World Development* 18: 605-615.

A review of successful diffusion cases involving renewable energy technologies. These include the Argentine windmill industry, the Nepali water turbine industry, and solar water heaters in various countries. These cases are followed by a detailed examination of different conditions that led to success in low-tech renewable energy diffusion. The diffusion framework consists of a demand for new technology, a manufacturer of new technologies (rural or urban, small-scale or large-scale), access to design information, and the distribution of technology. Certain types of renewable technology may not develop without government intervention. Describes types of government intervention, including the provision of information to customers and manufacturers, implementation of taxes and subsidies and credit services, and direct participation in equipment manufacture.

385. Hyman, Eric L. 1987. "The Strategy of Production and Distribution of Improved Charcoal Stoves in Kenya." *World Development* 15: 375-86.

Spread of charcoal stoves successful because: 1) built on mix of traditional designs and designs in other LDCs; 2) allowed evolution of stove through extensive field testing; 3) emphasized private-sector distribution.

386. Jayasuriya, S.K. 1984. "Technical Change and Revival of the Burmese Rice Industry." *Developing Economies* 22: 137-54.

Analyzes changes in the Burmese rice industry over the previous 20 years, in which Burma moved from a backstage player on the world rice market to a leader because of technological changes in production. Historical background, including early developments, postwar developments, and the era of revolutionary government. Looks at government investment policies. Holds that in the 1960s, the Burmese government emphasized the development of a domestic industrial sector based on heavy industries, which led to a decline in rice exports that had serious effects on the economy. In the early 1970s, the government began to emphasize exports and to encourage sectors with export potential, including agriculture. In the mid 1970s a production program for rice was launched based on a package of new rice technology; farmers were encouraged to adopt new technology. Data on the economics of the new technology are given.

387. Kaplinsky, Raphael. 1984. "Indigenous Technical Change: What We Can Learn from Sugar Processing." ***World Development*** **12: 419-32.**

Looks at the extent of local technological capacity in peripheral countries, with a focus on sugar processing. Charts the impact that recent technological improvements have had in changing the private and social desirability of choice. Finds that the development and maturation of this type of technology is similar to the pattern in other sectors, since protection of the state is important during the sector's long infancy. But finds that in sugar technology almost all of the effective developments occurred in the technology system, rather than within firms.

388. Katz, Jorge M. 1984. "Domestic Technological Innovations and Dynamic Comparative Advantage: Further Reflections on a Comparative Case-Study Program." ***Journal of Development Economics*** **16: 13-38.**

Analyzes factors influencing the speed of acquisition of new technologies. Uses the results of a survey of historical experience in Latin American industrial plants to examine (1) the sequence in which different types of capability are acquired, (2) the differing impact on acquisition of different ways of organizing production, (3) the effects of firm-level and wider macroeconomic variables on the nature and pace of acquisition, and (4) the effects of promoting technological capability acquisition by protecting infant industries in different circumstances.

389. Katz, Jorge M. (ed.). 1987. *Technology Generation in Latin American Manufacturing Industries*. New York: St. Martin's Press.

Important leading article, "Domestic Technology Generation in LDCs: A Review of Research Findings." This covers the work of a set of studies by the Latin American Research Programme on Science and Technology. Shows that significant technical change processes occur within limited sectors and in certain LDCs.

390. Kim, Linsu and James M. Utterback. 1983. "The Evolution of Organizational Structure and Technology in a Developing Country." *Management Science* 29: 1185-97.

Examines the evolutionary pattern of relationships among technology, structure, environment, and other contextual variables in 31 Korean manufacturing organizations. Results of bivariate analyses show that younger firms exhibit more mechanistic structure, a higher degree of adaptability in operations technology, a lower degree of adaptability in indigenous technical capability, a lower degree of innovation, and smaller scale than older firms. Younger firms also see suppliers as more important; older firms see customers and competitors as more important. The conclusions are supported by multivariate analysis.

391. Kiyokawa, Yukihiko. 1983. "Technical Adaptations and Managerial Resources in India: A Study of the Experience of the Cotton Textile Industry from a Comparative Viewpoint." *Developing Economies* 21: 97-133.

Addresses the question of why the Indian cotton textile industry lost its competitive power in the international market. Finds that the introduction of new technologies was rather quick because of the connection with British industries. Subsequent diffusion was slow because of regional differences and managerial defects. These causes are supported by regression and loglinear analysis. British technicians were important in choice of technique; this retarded diffusion of Ring frames. These technicians favored old, familiar technology based on experience in Britain. Diffusion of these frames was also hindered by the small size of depreciation funds.

392. Kiyokawa, Yukihiko. 1984. "Entrepreneurship and Innovations in Japan: An Implication of the Experience of

Technological Development in the Textile Industry." *Developing Economies* 22: 211-36.

Historical analysis of the rise of Japan's textile industry, including a section on the role of local technological innovations. Maintains that economic historians have generally explained Japan's rapid industrialization by reference to human resource factors, such as the early establishment of an educational system, adequate supply of disciplined labor, and flexible social value systems. The author wishes to revise this view by maintaining that entrepreneurship was the most significant factor in the promotion of industrialization. The first section reviews existing studies of Japanese entrepreneurs. The second looks at major innovations in textile technology and their promoters. The last section looks at implications of innovating activities for entrepreneurship.

393. Koppel, Bruce. 1980. "The Green Revolution as Social Development: The Case of the Philippines." *International Journal of Contemporary Sociology* 17: 82-113.

Offers two hypotheses about Green Revolution agricultural development strategies: (1) without technologies that offer the possibility of significant output or economic gains, little agricultural contribution to local market expansion will be made; (2) the time required to introduce new production technologies can be shortened through effective use of compensatory or "re-equilibrating" policies. Then tests these hypotheses with data regarding rice farming from 338 Philippine barrios. Finds that the "best" strategy to diffuse a technology is not the same as the best technology to get it utilized; technology must be "indigenized" (i.e., integrated into local processes) before it really spreads.

394. Lall, Sanjaya. 1980. "Developing Countries as Exporter of Industrial Technology." *Research Policy* 9: 24-53.

Despite the prevailing view that technological progress is unlikely in LDCs, some semi-industrialized LDCs are showing that they can compete in international technological markets. Offers specific discussions of India, Brazil, Argentina, Mexico, Korea, Taiwan and Hong Kong. This competitiveness is most likely in industries where pace of change is slow and large R&D expenditures are unnecessary.

395. Lee, Jinjoo, Zong-tae Bae and Dong-kyu Choi. 1988. "Technology Development Processes: A Model for a Developing Country with a Global Perspective." ***R&D Management*** **18: 235-50.**

Presents a conceptual model for development processes in LDCs. Proposes three development stages: initiation, internalization and generation. These stages correspond to the fluid, transition and specific stages of the Utterback and Abernathy dynamic model of development, but they represent an attempt to extend the model from developed to developing countries. In addition, this new model is said to include firms with continuous flow production processes, and while the unit of analysis in the previous model was limited to the industry level, this model extends it to the corporate level.

The model is based on two assumptions: (1) Most new technologies come from developed countries, are transferred to LDCs by various channels and are finally developed and mastered in LDCs. (2) Technological development in LDCs follows macro-level stages such as initiation, internalization, and generation at the national, industrial, and firm levels. Proposes that old and mature technologies are mainly transferred by informal channels and new and high technology by formal channels. The paper contains useful tables dealing with theories of tech development, including a table summarizing studies on technology development processes in LDCs, from Enos in 1962 to Lall in 1980.

396. Levy, Brian. 1988. "The State-Owned Enterprise as an Entrepreneurial Substitute in Developing Countries: The Case of Nitrogen Fertilizer." ***World Development*** **16: 1199-1211.**

Traces the development of the industry in India, and shows how state investment can serve as a replacement for private investment. The article is based on the concept of variations in the conditions of entry across different sectors of industry. Demonstrates how these variations across sectors suggest a role for state-owned enterprises in developing countries. Investigates the degree to which the entrepreneurial substitution hypothesis can account for the dominant role of state-owned enterprises in the nitrogen fertilizer industry. Offers a good historical summary of the background of international organization in the nitrogen fertilizer industry, and of the evolution of ownership in the ammonia industry in India, Brazil, the Republic of Korea, and Mexico.

397. Lichtman, Rob. 1987. "Toward the Diffusion of Rural Energy Technologies: Some Lessons from the Indian Biogas Program." ***World Development*** **15: 347-74.**

Shows the importance of basing technology diffusion efforts on an understanding of rural resource flows and local political economy. Discusses conflict between technology-focused and people-focused approaches.

398. Mahnken, T.G. and T.D. Hoyt. 1990. "The Spread of Missile Technology to the Third World." ***Comparative Strategy*** **9: 245-63.**

Unstable regions in the developing world are increasingly acquiring the technology to field more lethal ballistic missiles with longer range and higher accuracy. This suggests a broader proliferation policy that moves beyond supply-side efforts to restrict technology to efforts to decrease demand for and utility of ballistic missile systems.

399. McDonald, Ronald H. and Nina Tamrowski. 1987. "Technology and Armed Conflict in Central America." ***Journal of Interamerican Studies and World Affairs*** **29: 93-108.**

Argues that technological change has increased both armed conflict and revolutionary activity, as well as their likelihood of success, in Central America.

400. Morehouse, Ward. 1982. "Opening Pandora's Box: Technology and Social Performance in the Indian Tractor Industry." Pp. 180-208 in ***The Economics of New Technology in Developing Countries*****, edited by Frances Stewart and Jeffrey James. Boulder, CO: Westview.**

Traces the development of the Swaraj tractor in India, a local technology which has become competitive with foreign sources, though requiring protection in early phases.

401. Pavitt, K. 1984. "Sectoral Patterns of Technical Change: Towards a Taxonomy and a Theory." ***Research Policy*** **13: 343-73.**

Study of 2,000 significant innovations in Britain since 1945. Develops a firm taxonomy (supplier dominated, production intensive, and science based) based upon source of technology, requirements of users, and potential for appropriation. Not specifically about Third World countries, but still an important source article for understanding innovation in firms because it argues most technological knowledge is not generally applicable and easily reproducible, but specific to firms and applications.

402. Piniero, Martin and Eduardo Trigo. 1983. "Public Policy and Technical Change in Latin American Agriculture." ***Food Policy*** **8: 46-66.**

Interprets the modernization and technical change of Latin American agriculture over the previous two decades. The innovation process is analyzed through eight case studies of different products in a variety of countries and with a variety of production methods. Evidence provided in the case studies suggests that inducement mechanisms were effective at the technology-generation level in cases of agrarian initiative. Most of the technology used for certain products was created in the developed countries. The authors suggest that in these circumstances the inducement mechanisms were unable to adjust the innovation process to the specific needs of the continent.

403. Pitt, M.M. and G. Sumodiningrat. 1991. "Risk, Schooling and the Choice of Seed Technology in Developing Countries: A Meta-Profit Approach." ***International Economic Review*** **32: 457-474.**

Uses a large-sample survey of Indonesian family households to examine determinants of seed variety choice. An econometric model is employed, involving a simultaneous equation switching regimes model with random profit functions. The adoption of high-yielding varieties is positively associated with profitability, likelihood of flooding, quality of irrigation conditional on relative profit, and availability of credit. Adoption of these varieties is negatively associated with the likelihood of drought and the amount of land owned. The authors remark that the reduced likelihood of HYV adoption by larger landowners is consistent with risk aversion increasing with wealth. Education is not found to be a significant determinant of variety choice.

404. Quiroga, Eduardo R. 1984. "Irrigation Planning to Transform Subsistence Agriculture: Lessons from El Salvador." ***Human Ecology*** **12: 183-202.**

A case study of the irrigation district of Zapotitan in El Salvador. Illustrates that the transformation of subsistence agriculture through irrigation demands new institutions designed not only to service agricultural production, but also to ensure the accrual of benefits to the targeted group.

405. Rauniyar, G.P. and F.M. Goode. 1992. "Technology Adoption on Small Farms." ***World Development*** **20: 275-82.**

Investigates interrelationships among technological practices adopted by maize-growing farmers in Swaziland. Considers seven practices (per hectare application of high-yielding variety seed, basal and topdress fertilizer, insecticides, tractor plowing, plant populations, average planting date) and uses factor analysis to examine relationships among them. Results show that they can be summarized by three distinct technological practices. Although technology adoption requires simultaneous decisions by farmers on the use of practices within a package, the three packages are independent of one another. Suggests that understanding of interrelationships among practices is important for successful technology planning in developed countries.

406. Reddy, N. Mohan, John D. Aram, Leonard H. Lynn. 1991. "The Institutional Domain of Technology Diffusion." ***Journal of Product Innovation Management*** **8: 295-304.**

Suggests that the institutional scope for understanding technology diffusion should include organizations that manufacture technological complementarities, institutions that possess vertical complementary assets, and the nonmarket sector. The nonmarket sector includes trade associations, professional societies, governmental agencies, independent research agencies, and public service organizations. Develops 16 propositions about these institutions, and discusses the implications of this framework for marketers of technical products.

407. Street, J.H. and Dilmus D. James. 1978. "Closing the Technological Gap in Latin America." ***Journal of Economic Issues*** **12: 477-500.**

Maintains that the technological gap widened as a result of the 1973 rise in oil prices. A more effective strategy for the stimulation of science and technology is needed. Gives cases of recent progress in this direction in the Latin cultural environment.

408. Szyliowicz, Joseph S. 1981. "Technology, the Nation State." Pp. 1-39 in *Technology and International Affairs*, edited by Joseph S. Szyliowicz. New York: Praeger.

An overview of the articles in this volume. Outlines two basic models for the overall innovation process--"discovery push" and "demand pull." Demand pull model accounts for about 70% of all innovations, but these innovations tend to be minor rather than breakthroughs. Discovery push (supply-side) has been the implicit basis of most scientific and technological policies. Efforts of developing countries in this area have seldom yielded significant results. Discusses the transition from an administrative (Weberian bureaucratic) to scientific state, based on scientific rather than legal rationality and organized by "decentralized functional networks."

409. Teitel, Simon. 1981. "Creation of Technology Within Latin America." *Annals of the American Academy of Political and Social Science* 458: 136-50.

Important early statement of the view that, contrary to the traditional assumption of the literature on technology transfer, significant innovation does take place in the Third World. Available evidence based on enterprise and industry studies in Latin America indicates the existence of a fair amount of technological creativity. Suggests that study is needed to understand more fully the processes of technology transfer, adaptation, and creation in semi-industrialized countries.

410. Teitel, Simon. 1984. "Technology Creation in Semi-Industrial Economies." *Journal of Development Economics* 16: 39-62.

Defines industrial technology as the technical and organizational information required to manufacture industrial products. Technical change refers to all modifications of such information. Discusses aspects of the interaction between observed technical change and the economic environment in semi-industrialized countries in Latin America. The article then discusses technical change in these countries and its possible conceptualization in the light of the market failure metaphor. Looks at the relationship between stages of technical development and the technical information and skill requirements for manufacturing industries. A major conclusion: Technical change in Latin America has mainly taken the form of adaptation of imported technologies to local environments, rather than cost-reducing improvements and creation of new technologies. No broad generalizations are valid regarding the effect of industrial promotion policies on technical change.

411. Teubal, Morris. 1984. "The Role of Technological Learning in the Exports of Manufactured Goods: The Case of Selected Capital Goods in Brazil." *World Development* 12: 849-65.

Measures the economic and export benefit of technological learning within firms, and attempts to gauge how much technological learning shows up in actual products.

412. Wunsch, James S. 1977. "Traditional Authorities, Innovation, and Development Policy." *Journal of Developing Areas* 11: 357-72.

Looks at the role of traditional authorities in technological development, drawing upon a year of field research in two Ghanaian cities, backed by approximately 150 interviews. Finds that traditional leaders often play a crucial role in legitimizing development programs and expediting their execution. The importance of leaders varies based upon environmental factors--for example the viability of traditional authorities--and program-specific factors--for example, whether modification of conventional modes of behavior or coordinated community action is required.

413. Yapa, Lakshman S. 1980. "Diffusion, Development, and Ecopolitical Economy." Pp. 101-41 in *Innovation Research and Public Policy*, edited by John A. Agnew. Syracuse: Syracuse University Press.

In the standard description of underdeveloped states, the economy has two sectors (modern, traditional). From this point of view, eradicating poverty is largely a matter of improving productive forces. But early years of Green Revolution gave the lie to the alleged traditionalism of peasants. MNCs have capital-intensive production methods, destroy jobs in poor countries, and spend large sums advertising non-essential goods. Ecopolitical economy: begins with Marxian concept of mode of production, but the concept is expanded to include modern relations of consumption, ecology, and geology. Poverty is due to the position of countries in the international economy. Green Revolution shows insufficient attention to social factors.

Chapter 6 -- Agriculture

414. Adams, John and Balu Bumb. 1979. "Determinants of Agricultural Productivity in Rajasthan, India: The Impact of Inputs, Technology, and Context on Land Productivity." ***Economic Development and Cultural Change*** **27: 705-22.**

Investigates the sources of variation in agricultural productivity, defined as the average value of crop output in the districts in Rajasthan. Land productivity depends on three things: supplies of conventional inputs, cropping pattern and intensity, and the use of modern mechanical and chemical technologies. Infrastructure and institutions are found to be facilitative or intermediary variables.

415. Alauddin, Mohammad and Clem Tisdell. 1988. "Has the Green Revolution Destabilized Food Production?: Some Evidence from Bangladesh." ***Developing Economies*** **26: 141-60.**

Analyzes data on the variability of Bangladesh foodgrain production and considers the possible role of Green Revolution technology in moderating or accentuating fluctuations in production and yield. Concludes that the Green Revolution may have resulted in a reduction of relative variability of agricultural production; the probability of a particularly low yield may have been reduced.

416. Alauddin, Mohammad and Clem Tisdell. 1988. "Impact of New Agricultural Technology on the Instability of Foodgrain Production and Yield: Data Analysis for Bangladesh and Its Districts." ***Journal of Development Economics*** **29: 199-228.**

Uses data from Bangladesh that indicates that the relative variety of foodgrain production and yield has fallen with the adoption of Green Revolution technologies. Pages 201-203 offer a reasonably good literature review of works on stability and adaptability of yields and of recent in-depth studies of Indian agriculture.

417. Alauddin, Mohammad and Clem Tisdell. 1989. "Poverty, Resource Distribution and Security: The Impact of New Agricultural Technology in Rural Bangladesh." ***Journal of Development Studies***

25: 550-70.

Offers prodigious evidence that Bangladesh is experiencing a growing concentration of land, agricultural technology, and ancillary resources. The burgeoning landlessness produces greater dependence on wage employment for rural subsistence. Authors argue that "non-exchange income" grows in importance during slack years, especially "abnormal" years when both wages and employment fall--a "cushioning effect" that is being undermined by population growth, depletion of natural resources, and penetration by technological/market forces.

418. Anderson, R.S., P.R. Brass, E. Levy, and B.M. Morrison (eds.). 1982. *Science, Politics, and the Agricultural Revolution.* Boulder: Westview Press.

Includes articles on the commercialization of Indian agriculture through technology transfer and discussions of possible "routes to agrarian changes." Two articles focus on Indian higher education; others emphasize the problems of transfer and research as applied to agriculture.

419. Bayri, Tulay Y. and W. Hartley Furtan. 1989. "The Impact of New Wheat Technology on Income Distribution: A Green Revolution Case Study, Turkey, 1960-1983." *Economic Development and Cultural Change* 38: 113-28.

Disputes the perception that High-Yield Variety wheat hurts laborers. Research shows real wages dropping when HYV crops were introduced mostly because population growth outstripped gains induced by these agricultural improvements. However HYV rural strategies were somewhat labor saving, contrary to expectation, only because wheat is less labor intensive than the crops it replaced.

420. Biswas, Margaret R. 1979. "Nutrition and Agricultural Development in Africa." *International Journal of Environmental Studies* 13: 207-17.

Problem of severe malnutrition in Africa is particularly acute among young children and pregnant women. Various disorders are discussed, together with the causes: seasonal food shortages, maldistribution, ignorance (especially with respect to weaning), urbanization.

421. Blyn, George. 1983. "The Green Revolution Revisited." ***Economic Development and Cultural Change*** **31: 705-26.**

Report on research carried out in Punjab and Haryana to review findings of a paper originally published by the author in 1979. Revises his earlier findings to suggest that tractorization provides for social betterment.

422. Bowonder, B. 1979. "Impact Analysis of the Green Revolution in India." ***Technological Forecasting and Social Change*** **15: 297-314.**

A detailed analysis of the Green Revolution's impact on India's farmers, concluding that poor farmers have become poorer and that government assistance is required to repair the damage done.

423. Burke, Robert V. 1979. "Green Revolution Technologies and Farm Class in Mexico." ***Economic Development and Cultural Change*** **28: 135-54.**

Reports results of an empirical investigation into Green Revolution technologies in Mexico, asking whether they favor large farmers or small farmers. Finds that substantially more large farms were classified as using the new technologies than small farms.

424. Buttel, Frederick H., Martin Kenney and Jack Kloppenburg, Jr. 1985. "From Green Revolution to Biorevolution: Some Observations on the Changing Technological Bases of Economic Transformations in the Third World." ***Economic Development and Cultural Change*** **34: 31-56.**

Argues that the major characteristics of the emergent Biorevolution can be most clearly gauged by comparison with the experiences of the Green Revolution. Dependency is just as likely to result from biotechnology.

425. Clemens, Harrie and Jan P. De Groot. 1988. "Agrarian Labor Market and Technology Under Different Regimes: A Comparison of Cuba and the Dominican Republic." ***Latin American Perspectives*** **15: 6-36.**

Compares the development of technologies used in sugar cane harvest in the Dominican Republic and Cuba. The first is taken as an example of a capitalist peripheral economy and the second as an example of a socialist peripheral economy. Concludes that labor market policy in the socialist country leads to a tightening of the labor market and that of the capitalist country leads to keeping an ample labor supply. This has made Cuba more willing to adopt labor-saving devices and has led to a greater mechanization of its agricultural sector. Good references on sugar production in these countries, in English, Dutch, and Spanish.

426. Conley, Dennis M. and Earl O. Heady. 1981. "The Interregional Impacts of Improved Truck Transportation on Farm Income from Rice in Thailand." *Journal of Developing Areas* 15: 549-60.

Analyzes improved transportation technologies as a means of increasing wealth and sectoral development for the peasantry. Outlines a model with which the impact of a potential national transportation improvement can be evaluated to determine its effects within regions of a country, and applies this method to improvements in Thailand's truck transportation.

427. Coxhead, Ian A. and Peter G. Warr. 1991. "Technical Change, Land Quality, and Income Distribution: A General Equilibrium Analysis." *American Journal of Agricultural Economics* 73(2): 345-360.

Using data from the Philippines, examines the short-run income distributional effects of technical change in agriculture, by means of a small equilibrium model. This model distinguishes between commodities that are traded internationally and those that are not, between fixed and mobile factors of production, and between technical changes of different factor bases. It also distinguishes between groups of agricultural producers with and without access to resources, such as irrigation, that are complementary with the new technology. The interests of owners of fixed factors of production do not necessarily coincide. The distributional changes produced by technical change are highly sensitive to factor biases and to prior differences in land quality.

428. Diwan, Romesh and Renu Kallianpur. 1985. "Biological Technology and Land Productivity: Fertilizers and Food Production in India." *World Development* 13: 627-38.

Determines that technological change dealing with the biological side of production, especially the introduction of fertilizers, has little impact on grain output.

429. Farmer, B.H. 1979. "The 'Green Revolution' in South Asian Ricefields: Environment and Production." *Journal of Development Studies* 15: 304-19.

Emphasizes the climatic/environmental factors determining the rejection or acceptance of technology.

430. Flinn, J.C. and P.B.R. Hazell. 1988. "Production Instability and Modern Rice Technology: A Philippine Case Study." *Developing Economies* 26: 34-50.

Analyzes the effects of new production technologies on the industry, particularly on variety and stability of the crops. Concludes that growth in rice production in the Philippines has been dominated by a yield as opposed to an area effect. Relative production variability has changed little, but modern rice technology does appear to have increased absolute production variability. Increases in dry-season rice have reduced production variability and increases in wet-season rice have led to greater variability.

431. Flora, Cornelia Butler and Jan L. Flora. 1989. "An Historical Perspective on Institutional Transfer." Pp. 7-31 in *The Transformation of International Agricultural Research and Development*, edited by J. Lin Compton. Boulder: Lynne Rienner Publishers.

Uses history of U.S. agricultural institutions to understand problems of establishing agricultural institutions in developing countries. Offers a history of the U.S. land-grant system, suggesting that agrarian technological changes were closely linked to urban-industrial transformation. Hatch and Smith-Lever Acts are identified as central funding mechanisms for producing innovations in U.S. agriculture. Post-WWII attempts to export the land-grant system, under the Point IV Program and succeeding U.S. aid programs. Suggests that human capital has been created in developing countries, research and extension units will need to be finely tuned to local needs and will need local political support to serve the limited-resource farmer in order to maintain the human capital. New agricultural institutions will have to use Western technology selectively and they will have to achieve their ends more rapidly than was the case for U.S.

land-grant institutions.

432. Flores-Moya, Piedad, Robert E. Evenson and Yujiro Hayami. 1978. "Social Returns to Rice Research in the Philippines: Domestic Benefits and Foreign Spillover." ***Economic Development and Cultural Change*** **26: 591-608.**

Presents a formal model that analyzes the costs and benefits of rice research in the Philippines. Finds that in the closed- economy case the consumers are the sole beneficiaries of research; the producers are made worse off. In the open-economy case, producers are better off and consumers continue to enjoy the same level of economic welfare, while foreign exchange is saved.

433. Fried, Jacob, Roberto Armijo and Manuel Trejo. 1984. "Assessing Techno-Economic Alternatives for Rural Development in Northeast Mexico." ***Technological Forecasting and Social Change*** **25: 61-81.**

Assesses means of choosing production methods compatible with resource level and environment. Shows how recent advances in computer mapping, linear programming, and techno-social modelling can help planners and decision makers. Finds that cooperatives are not a very good way of exploiting natural resources or of maintaining ecological balance between communities and resources, due to variability of community cohesiveness and of organizational requirements of different techniques.

434. Goody, Jack. 1980. "Rice-Burning and the Green Revolution in Northern Ghana." ***Journal of Development Studies*** **16: 136-55.**

Discusses the dramatic social changes brought about by the Green Revolution, and the subsequent violent reaction to the changes. Maintains that the introduction of HYVs into northern Ghana has led to the mechanized farming of rice in the river valleys. Before, this land was used only occasionally as pasture. Afterwards it was cultivated by civil servants, military men, and others who were encouraged by government policy and by desire for profit. The system of land tenure in these areas has been radically changed by the introduction of these large farms, since written evidence of ownership is needed for loans and in cases of land dispute. The conflicts over land, and the attendant increase in social differentiation, have led to the burning of rice and the destruction of machinery.

435. Greeley, Martin. 1986. "Rural Energy Technology Assessment: A Sri Lankan Case Study." *World Development* 14: 1411-21.

Windmill technology was clearly inferior economically to fossil-fuel technology. Findings based on the results of a comparative survey on costs and benefits of windmills and kerosene pumps for irrigation of subsidiary food crops in the Dry Zone of Sri Lanka. Farmers were using windmills intensively but they often had to use kerosene pumpsets. Windmills were generally cheaper. Windmill farmers cultivate smaller areas and get lower average yields.

436. Griffin, Keith. 1974. *The Political Economy of Agrarian Change*. Cambridge: Harvard University Press.

A major, frequently cited critique of the Green Revolution. Argues that the new seed varieties, with their heavier inputs of fertilizer, served to increase inequalities in the developing countries involved.

437. Herdt, Robert W. 1987. "A Retrospective View of Technological and Other Changes in Philippine Rice Farming, 1965-1982." *Economic Development and Cultural Change* 35: 329-50.

Attempts to document some changes in Philippine rice farming at the farm level and measure their impact on direct participants in the production process. Examines these questions: Have incomes of farmers increased significantly? Have laborers' incomes from rice work declined or increased? Has land reform made a difference to farmers' income from rice? What is the impact of irrigation on farm income? Who retains the benefit of various changes? Many of the findings are inconclusive, however. Real incomes of farmers and agricultural laborers have shown no dramatic changes despite the substantial increase in production, but real rice prices have declined, permitting consumers to purchase their rice at a lower real cost.

438. James, Jeffrey. 1989. *Improving Traditional Rural Technologies*. London: Macmillan.

Offers a "micro-procedure" for the design and development of AT techniques--intended to speed up how quickly these ATs can be replicated across communities. The procedure includes a means to assess the likely impact of technology introduction on the poor, an emphasis on a "learning process approach," and coordination so that modernization would occur over the entire rural sector.

439. Juma, Calestous. 1989. *The Gene Hunters: Biotechnology and the Scramble for Seeds*. Princeton, NJ: Princeton University Press.

History of biotechnology and current issues in the field presented in the context of the relation between agricultural growth and the European colonization of Africa. Argues that this colonization can only be understood in light of competition for genetic resources, such as new plant varieties. Deals with historical botany and the role of genetic material in socio-economic development. African countries need sound policies on genetic resource conservation and biotechnology in order to take advantage of emerging technologies. Offers a case study of germplasm and agriculture in Kenya. Good bibliography of works on present-day biotechnology and historical botany.

440. Junne, Gerd. 1987. "Biotechnology: The Will to Manipulate Is Human. But What Do We Want?" *Development* 29: 3-90.

Special issue under the guest editorship of Gerd Junne (University of Amsterdam). Consists of: (1) Introduction, general remarks and policy options for dealing with the problem of declining prices that results from the increased production of biotechnology. (2) Biorevolution: Myth or Reality. David Dembo et al. ("Biotechnology and the Third World: Caveat Emptor") argue that biotechnology has several characteristics that make it readily adaptable to the Third World, including lower scientific barriers to entry than other technologies, location specificity of products and inputs, increased bargaining power of developing countries with TNCs for products related to biotech, lower labor costs leading to lower R&D costs in developing countries, lower investments per unit of production, and the relatively small scale of factories. Robert W. Herdt ("Equity Considerations in Setting Priorities in Third World Rice Biotechnology Research") uses economic theory to prioritize research problems in rice biotech. (3) Examples of biotechnology, including case studies and studies of specific issues (consequences of biotech for West African cocoa smallholders, biotech and natural sweeteners, vanilla and biotech, seed industry development in North-South perspective, the scope for cooperation between the U.S. and Mexico in biotech for food, and biotech in India). (4) Research, Priorities, and Networks. Contains a variety of articles on social impacts, the role of the university, policy research, the role of international organizations, biotech assessment, a global network of microbial research centers, and future social science research on the social implications of biotech.

441. Kaimowitz, David. 1990. "Placing Agricultural Research and Technology in One Organization: Two Experiences from Colombia." ***Public Administration and Development*** **10: 199-208.**

Finds that placing agricultural research and technology transfer in the same institution is neither a necessary nor a sufficient condition for effective coordination between the two activities. Other factors listed include: (1) specificity of problems addressed; (2) institution's capacity to manage coordination; (3) status differences and competition over resources; (4) institutional size; (5) level of politicization of technology transfer activities.

442. Kenney, Martin and Frederick Buttel. 1985. "Biotechnology: Prospects and Dilemmas for Third World Development." ***Development and Change*** **16: 61-92.**

Attempts to explore the constraints that developing countries face and the strategies that developing countries must adopt in order to meet the challenges and opportunities afforded by biotechnology. Describes institutions in the developed countries that have been central in making applied molecular and cell biology a reality. Describes implications for developing countries of developmental directions being chosen by these institutions. Briefly explores possible responses by developing countries and institutions to a trajectory of biotechnology in developed countries that the authors believe has limitations in meeting needs of developing countries.

443. Koppel, Bruce and Edmund Oasa. 1987. "Induced Innovation Theory and Asia's Green Revolution: A Case Study of an Ideology of Neutrality." ***Development and Change*** **18: 29-67.**

Begins by reviewing the "neutrality claim" of contemporary agricultural research. This claim, they maintain, absolves the agricultural research system from accountability from the socio-economic consequences of its technological choices. Describes the influential "induced innovation" theory, as applied to agriculture by

Yujiro Hayami and Vernon Ruttan, then looks at the development of the theory. Concludes by examining induced innovation theory as ideology by analyzing the theory's evolution and practice.

444. Leaf, Murray J. 1983. "The Green Revolution and Cultural Change in a Punjab Village: 1965-1978." *Economic Development and Cultural Change* 31: 227-70.

Describes the main outlines of the Green Revolution in one village in Punjab between 1965 and 1978. Concludes that the gains have been real in both ecological and economic terms, despite inflation, and that they have been accompanied by increased equity and political stability.

445. Lipton, Michael. 1988. "The Place of Agricultural Research in the Development of Sub-Saharan Africa." *World Development* 16: 1231-57.

Superb article asking why agricultural research in Sub-Saharan Africa costs more and produces less. The small size of countries and research organizations, their dispersion, and high turnover make it hard to attain a critical mass of research scientists. Researchers have failed to concentrate on important crops and export crops. European biases and low funds worsen the problem.

446. Pearse, Andrew. 1977. "Technology and Peasant Production: Reflections on a Global Study." *Development and Change* 8: 125-59.

Author was Project Manager of the UNRISD/UNDP research project and a noted critic of the Green Revolution. This article studies three related aspects of technological change in the agricultural/rural sector: (1) identification of factors obstructing or facilitating acquisition and use of technology based on specially bred plant varieties and use of manufactured chemical compositions for plant nutrition, (2) identification of economic and social changes which follow large-scale introduction of technology, and (3) assessment of measures and programs proposed and carried out by governments in order to manage or modify processes set off by technological change. Sees the effect of the new technology as further pauperization of peasantry. Suggests that much is to be learned from Chinese strategies of controlling socio-economic effects of new technology.

447. Perrolle, Judith A. 1985. "Building a Better Goat: The Prospects for an Indonesian Development Project." *International Journal of Contemporary Sociology* 22: 241-64.

This is an assessment of a Green Revolution program, from the AT perspective. After reviewing development priorities for the introduction of small ruminant production in Indonesia, especially Java, evaluates the potential of two alternative approaches. Suggests that an "intermediate level goat"--neither individual family nor commercially produced--based upon village-cooperative production might be the best compromise, although problems of appropriateness, division of income, utilization of products, and nutrition would still be problems.

448. Platteau, Jean-Philippe. 1984. "The Drive Towards Mechanization of Small-Scale Fisheries in Kerala: A Study of the Transformation Process of Traditional Village Societies." *Development and Change* 15: 65-103.

Shows how technology has disrupted traditional patterns of living. The purpose of the paper is to illustrate the rise and development of the market economy in the case of a fishing village in South Kerala (India) open to modernization forces. Analyzes three types of change in this village: (1) the increasing complexity of the marketing process, (2) the polarization of the socio-economic structure, (3) the opening up and de-linking of product and factor markets.

449. Plucknett, Donald L., Nigel J.H. Smith, Selcuk Ozgediz. 1990. *Networking in International Agricultural Research*. Ithaca, NY: Cornell University Press.

Discussion of the philosophy and practice of "networking" to promote agriculture research, using examples from international agricultural research centers and national institutions. Offers a typology, conceptual framework, and model of developmental stages before examining particular types of networks: information exchange, material exchange, scientific consultation, and collaborative research networks.

450. Ridler, Neil B. 1983. "Labour Force and Land Distributional Effects of Agricultural Technology: A Case Study of Coffee." *World Development* 11(7): 593-9.

Studies the effect of new coffee technology on the agricultural labor market by investigating land reform and the effects it would have on coffee crops of Colombia. Argues that while new techniques of coffee cultivation will increase output-labor, output-land, and labor-land ratios, implying an increasing coffee employment and output, economic and institutional barriers will preclude higher output. These developments would threaten coffee employment and small coffee producers, resulting in land consolidation and labor displacement.

451. Rigg, Jonathan. 1989. "The New Rice Technology and Agrarian Change: Guilt By Association." ***Progress in Human Geography*** **13: 374-400.**

Argues that the biochemical aspect of the Green Revolution does not necessarily require concomitant dependency-inducing mechanization, although the two have been associated when the Green Revolution is analyzed. Also questions whether the expansion of these technologies is tied to growing inequalities observed in village life and the rural economy.

452. Rogers, Everett M. 1989. "Evolution and Transfer of the U.S. Extension Model." Pp. 137-152 in ***The Transformation of International Agricultural Research and Development*****, edited by J. Lin Compton. Boulder & London: Lynne Rienner Publishers.**

Maintains that the agricultural extension model in the United States was transformed since 1911 relatively effectively for domestic use, but that the model was transferred to Third World nations without adequate modification. This was mainly because the latter could not afford a sufficient number of extension workers and because agricultural research was not effectively connected with extension. The author suggests that both social and agricultural change are necessary to technology transfer.

453. Ruttan, Vernon W. 1989. "The International Agricultural Research System." Pp. 173-205 in ***The Transformation of International Agricultural Research and Development*****, edited by J. Lin Compton. Boulder & London: Lynne Rienner Publishers.**

Places the evolution of the international research system in the context of other post-WWII agricultural development assistance programs, and discusses organizational and management issues in the international agricultural research system. Contains a table of international agricultural research institutes, giving location, nature of research, geographical coverage, date of initiation, and budgets for 1988.

454. Sachs, Carolyn and Virginia Caye. 1989. "Women in Agricultural Development." Pp. 85-111 in *The Transformation of International Agriculture and Development*, edited by J. Lin Compton. Boulder & London: Lynne Rienner Publishers.

Looks at the effect of the U.S. land-grant system model on the situation of women in developing countries. Maintains that women's participation in agriculture is extensive and becoming more visible and that transfer of the land-grant system at present both alleviates and contributes to women's problems. Increased data collection should answer the problem of inadequate documentation of women's work, but research findings on this topic must be interpreted to policy makers and translated into action. The authors suggest that the sexual stratification systems and bureaucratic structures limiting women's possibilities in U.S. and Third World agriculture should be revised, and that coordination of specialists in areas such as home economics, agriculture, nutrition, and family planning would be a positive development.

455. Ventura, Arnoldo K. 1982. "Biotechnologies and Their Implications for Third World Development." *Technology in Society* 4: 109-30.

Biotechnology is pertinent to LDCs because of their highly agricultural economies. It can speed food production, and thereby address direct hunger problems, as well as create employment. Suggests steps to improve developing countries' biotechnological positions.

456. Vessuri, Hebe M.C. 1980. "Technological Change As the Social Organization of Agricultural Production." *Current Anthropology* 21: 315-27.

Broad discussion of technological change in the agriculture of contemporary Latin American societies. Emphasizes that "production is a social activity," and that anthropology should concentrate on actors and the social relations they establish in the productive process. Followed by the comments of various scholars.

457. Von Braun, Joachim. 1988. "Effects of Technological Change in Agriculture on Food Consumption and Nutrition: Rice in a West African Setting." ***World Development*** **16: 1083-98.**

Technological change that brought increased income resulted only in higher caloric intake and better nutrition, not in capital investment. Increasing productivity in a "woman's crop" only increased the flow of men into production (rather than helping women farmers).

458. Vondal, Patricia J. 1987. "Intensification Through Diversified Resource Use: The Human Ecology of a Successful Agricultural Industry in Indonesian Borneo." ***Human Ecology*** **15: 27-52.**

Examines the success of commercial duck egg production in the swamplands of South Kalimantan through the utilization of a human ecology framework. Seasonality of resource availability and human population growth are identified as two major constraints to production faced by farmers.

459. Wade, Robert. 1979. "The Social Response to Irrigation: An Indian Case Study." ***Journal of Development Studies*** **16: 3-26.**

Anthropological case study of the social response to irrigation in a Southern Indian district's villages. Addresses the question of whether canal irrigation is an important determinant of forms of village organization, as a result of the need for collective management of this type of irrigation. Finds that a relationship exists between the degree of corporate organization and scarcity and uncertainty of water supply. However, a number of other factors, such as the nature of irrigation works, management of irrigation works, and involvement of government in water distribution, weakens the relationship. As a result of these factors, the author suggests that the hypothesis of a relationship between social organization and water scarcity would be less useful in East and Southeast Asia than in India.

460. Walker, Thomas S. and K.G. Kshirsagar. 1985. "The Village Impact of Machine-Threshing and Implications for Technology Development in the Semi-Arid Tropics of Peninsular India." ***Journal of Development Studies*** **21: 215-31.**

An intensive study of the introduction of machine threshing into one region of India. Finds that technology did not really help with output. Threshing did not significantly reduce producer costs, increase cropping intensity, or greatly harm unskilled labor (mainly because the threshers were not applied to the region's primary crop). The main beneficiaries of the change were owners of the financial capital to buy and rent out machines.

461. Warhurst, Alyson. 1991. "Metals Biotechnology for Developing Countries and Case Studies from the Andean Group, Chile, and Canada." *Resources Policy* 17: 54-68.

Examines emerging applications and socioeconomic aspects of metals biotechnology. Considers the role of training in Andean metals biotechnology development, capability development in Chile and Canadian metals biotechnology policy. Suggests four sets of policy implications: (1) Policy research is required in order to investigate the obstacles to and implications of potential adoption of bacterial leaching by less-developed countries. (2) Since the productivity effects of metals biotechnology are both labor- and capital-saving, this technology may provide developing countries with new ways of expanding their mineral imports when prices are low, or of reacting to unstable conditions, thus safeguarding employment. (3) This technology is new and, as efforts at training in the Andean Pact, Chile and Canada show, will require multidisciplinary training at the mine site, as well as in the lab. (4) Metals biotechnology offers environmentally sound alternatives to conventional mining activities.

462. Warren, D.M. 1989. "Linking Scientific and Indigenous Agricultural Systems." Pp. 153-170 in *The Transformation of International Agricultural Research and Development*, edited by J. Lin Compton. Boulder & London: Lynne Rienner Publishers.

Concerned with interactions between agricultural scientists, agricultural extension staff, and indigenous producers. Surveys reasons for an early lack of interest in indigenous agricultural knowledge and practices, why this interest has begun to develop, and how this interest can improve agriculture and the lives of indigenous producers.

463. Waugh, Robert K., Peter E. Hildebrand, and Chris O. Andrew. 1989. "Farming Systems Research and Extension." Pp. 207-226 in *The Transformation of International Agricultural Research and Development*, edited by J. Lin Compton. Boulder & London: Lynne Rienner Publishers.

Looks at the farming systems approach to research and extension. This approach, dominant today, is said to have developed from the integrated rural development projects of the late 1960s and early 1970s, which involved a recognition that rural development involves a large number of interrelated activities. Concludes that the appearance of persons with farming systems experience in administration of research and extension institutions will result in more responsive and efficient organizations.

464. Zarkovic, Milica. 1987. "Effects of Economic Growth and Technological Innovation on the Agricultural Labor Force in India." ***Studies in Comparative International Development*** **22: 103-20.**

Examines the impact of the Green Revolution in north India, in particular the size of the agricultural labor force and the male/female participation rates. Uses Indian census data from 1961, 1971, and 1981 to show that the rural labor force in the Punjab increased while labor was displaced in Haryana. Overall, male predominance in the agricultural sector resulted.

Chapter 7 -- Research and Development

465. Abraham, I. 1992. "India's Strategic Enclave: Civilian Scientists and Military Technologies." ***Armed Forces and Society*** **18: 231-52.**

Examines changes in Indian security and the entrance of new actors into core areas of this policy area. Finds that the Indian security complex has become a diversified set of establishments that have grown up around two models. The first is based on a top-down structure that is less suited to technological innovation and concentrates on licensed production and production for the civilian market. The second model is more a more flexible and project-oriented system that has shown an ability to produce both nuclear devices and ballistic missiles. The "strategic enclave" is a subset of the security complex that is characterized by the second model.

466. Adhikari, Kamini. 1987. "Science, Society and the Indian Transformation." ***Philosophy and Social Action*** **13: 33-56.**

Special issue on "Social Problems of Scientific Knowledge." Adhikari examines the process of scientific change in India by specifying its different organizational modes and searching for "the idea and institution of 'Western science'."

467. Altbach, Philip G. 1989. "Higher Education and Scientific Development: The Promise of Newly Industrializing Countries." Pp. 3-29 in ***Scientific Development and Higher Education: The Case of Newly Industrializing Nations*****, edited by Philip G. Altbach et al. New York: Praeger.**

Examines the role of higher education in scientific development for the NICs. Issues considered include the international knowledge system, the language issue, the scientific diaspora, foreign training, traditional universities, the research infrastructure, research in small scientific communities, the role of scientific research in the NICs, and the role of the universities. Concludes that the NICs are at a turning point in their development. Future economic development will depend on R&D and on the harnessing of technology. The NICs appear to be at a take-off point for science and R&D, despite considerable challenges.

468. Altbach, Philip G. et al. (eds.). 1989. *Scientific Development and Higher Education: The Case of Newly Industrializing Nations.* New York: Praeger.

Contains papers on the role of higher education in developing scientific capacity and university-based research and use made of research in Korea, Malaysia, Singapore, and Taiwan. Also interested in the ways in which indigenous research has developed in these countries, the nature of local scientific development, and how indigenous science is related to international knowledge networks. The papers are mainly case studies, with four papers on higher education and scientific development, case studies on the academic research environment in Malaysia, Korea, Singapore and Taiwan, and a paper on publication strategies of scientists in the NICs.

469. Anandakrishnan, M. and Hiroko Morita-Lou. 1988. "Indicators of Science and Technology for Development." Pp. 293-304 in *Science, Technology, and Development*, edited by Atul Wad. Boulder: Westview Press.

Discusses indicators of development that can be used in planning for science and technology. Points out that "unique" social and economic indicators do not exist; an indicator's appropriateness and how it can be used depends on the sector concerned. Indicators can be compared across time or among countries with certain levels of some other indicator, such as per acre productivity. Describes current trends in the use of S&T indicators, including manpower, expenditures, patents, and bibliometric indicators.

Gives brief descriptions of four methodologies for using indicators: conventional uses, convergent partial indicators, distributive analysis, and correspondence values. Concludes that presently available indicators are potentially usefully, but practically insufficient as a policy-making tool for developing countries. Suggests that, given the lack of valid data to use as conventional quantitative indicators, new types of indicators need to be developed based on both quantitative and qualitative information.

470. Argenti, G., C. Filgueira, and J. Sutz. 1990. "From

Standardization to Relevance and Back Again: Science and Technology Indicators in Small, Peripheral Countries." *World Development* 18: 1555-1568.

Discusses inadequacy of standard indicators to cope with specific characteristics of S&T in small underdeveloped countries. As an example, uses a study of Uruguay carried out between 1986 and 1988. Argues that in countries like Uruguay one must take into account the low degree of institutionalization and the high degree of informality in developing, applying, and interpreting indicators of S&T activity.

471. Arvanitis, Rigas and Yvon Chatelin. 1988. "National Scientific Strategies in Tropical Soil Sciences." *Social Studies of Science* 18: 113-46.

Compares national strategies, finding that Third World scientific production is usually much higher than traditionally estimated, especially regarding local needs.

472. Bhalla, A.S. and A.G. Fluitman. 1985. "Science and Technology Indicators and Socio-Economic Development." *World Development* 13: 177-90.

Reviews the use of science and technology indicators, particularly for planning and evaluating development. Exposes problems with currently available statistics. Attempt to disaggregate the concept of indigenous technological capacity, which must be related to specific socio-economic objectives. Experts in diverse disciplines should be used to assist development planning, identifying appropriate indicators as needed.

473. Bharol, Chowdhry Ram. 1989. "Problems of R&D Management in India." *R&D Management* 19: 335-42.

Problem of manpower planning in Indian R&D organizations. Sections cover: (1) the lack of literature on the subject, (2) sources of scientific manpower statistics (mostly UNESCO) and statistics on R&D employment in India as compared to Japan, West Germany, the USSR, and India, (3) procedures for recruitment in India (mostly through magazine advertisements), (4) unemployment of S&T personnel, (5) the brain drain, (6) the internal brain drain due to underemployment or misemployment, and (7) suggestions for manpower planning.

474. Blickenstaff, J. and M.J. Moravcsik. 1982. "Scientific Output in the Third World." *Scientometrics* 4: 135-69.

Looks at publication of scientific writing in the Third World. Plots the number of scientific authors for every year between 1971 and 1976, standardizing for population size. Forty-three graphs, divided into broad geographic areas (Latin America, Africa, the Middle East, and Asia). Finds that in some countries scientific output is negligible, with huge variation among output of countries in every geographical region. Countries with a small output have about as good a chance to grow as countries with a larger output. Good source of empirical information on the amount of scientific research done in developing countries and on the growth of research. No insights into comparative quality of research.

475. Blume, Stuart S. (ed.). 1977. *Perspectives in the Sociology of Science*. New York: John Wiley and Sons.

Among the essays are "Towards a Relevant Sociology of Science for India" by Rodhika Romasubban and "Contrary Meanings of Science" by Stephen C. Hill. Romasubban offers a Marxian analysis showing how organization of production relations influences the utilization of science and technology in India, looking in particular at the "Green Revolution." Hill discusses changes in the working of a Thai research organization adopting an Australian lab form. He criticizes the Western notion that scientific organization can be the same anywhere.

476. Braun, T., W. Glenzel, and A. Schubert. 1988. "The Newest Version of the Facts and Figures on Publication Output and Relative Citation Impact of 100 Countries, 1981-85." *Scientometrics* 13: 181-8.

Consists of 5 tables: (1) total publication output in all science fields combined, (2) percentage share in world publication output, all science fields combined, (3) observed citation rate per publication, all science fields combined, (4) expected citation rate, all science fields combined, and (5) relative citation rate, all science fields combined. Gives publications containing details and additional information on scientometric principles and procedures underlying the data.

477. Clark, Norman. 1980. "The Economic Behavior of Research

Institutions in Developing Countries--Some Methodological Points." ***Social Studies of Science*** **10: 75-93.**

Attributes many of the problems with research in developing countries to micro-level implementation rather than to grand planning, although at least one macro-level attribute also is blamed: the lack of effective links within the scientific establishment. Calls for more empirical information about the S&T sector in developing countries and for "preventive assessment" of S&T-based problems rather than more traditional, hindsight-oriented evaluations. See Moravcsik's responses in volume 12.

478. Cottrill, Charlotte A., Everett M. Rogers, and Tamsy Mills. 1989. "Co-citation Analysis of the Scientific Literature of Innovation Research Traditions: Diffusion of Innovations and Technology Transfer." ***Knowledge: Creation, Diffusion, Utilization*** **11: 181-208.**

Use of author co-citation analysis (bibliometric data) to show the way these two specialties are related in the period from 1966-72. In general, these literatures developed independently with little cross-referencing: tech-transfer with a "producer" orientation, while diffusion research has a "user" orientation. Highly cited documents are used as exemplars.

479. Cueto, Marcos. 1989. "Andean Biology in Peru: Scientific Styles on the Periphery." ***ISIS*** **80: 640-58.**

Historical discussion of the Institute of Andean Biology and its emergence at the center of the new field of high altitude studies. The author describes two different styles in Peruvian physiology at the Institute, personified by two leaders in the field. Concludes that Peruvian physiology was not characterized by features of peripherality because: (1) qualified scientists with leadership qualities were on-site, (2) high altitude studies were a totally new area of research, (3) Peruvians had natural lab conditions without resorting to high technology, (4) the research specialty accorded with national interests, and (5) the specialty coincided with interests of other nations, so that the Peruvians could attract international funding.

480. Cueto, Marcos. 1990. "The Rockefeller Foundation's Medical Policy and Scientific Research in Latin America: The Case of

Physiology." *Social Studies of Science* 20: 229-54.

Four countries received large grants to spread U.S. academic model of medical schools, but failed to modernize Latin American research; the response was local resistance and conflict rather than passivity and imitation.

481. Davis, C.H. 1983. "Institutional Sectors of 'Mainstream' Science Production in Sub-Saharan Africa, 1970-79: A Quantitative Analysis." *Scientometrics* 5: 163-75.

Focusing on 36 Sub-Saharan African countries, shows the growing importance of universities, where 65 percent of the research originates.

482. Davis, C.H. and T.O. Eisemon. 1989. "Mainstream and Non-Mainstream Scientific Literature in Four Peripheral Asian Scientific Communities." *Scientometrics* 15: 215-39.

Describes mainstream output of scientific communities in Malaysia, Singapore, South Korea, and Taiwan and considers how well this output represents local activities in biochemistry, biology, physics, electrical engineering, and computer science. Also examines non-mainstream scientific literature in these four countries. Tables include number and percent of mainstream scientific publications by field in each country, aggregate characteristics of non-mainstream articles and references, proportion of non-mainstream scientific authors publishing in mainstream journals, degree of parochialism of non-mainstream literature by country and field.

Reaches two principal conclusions. First, in recent years three of these countries have experienced an explosion of mainstream scientific research and the fourth has enjoyed modest expansion. Yet the overall contribution of these countries to mainstream science is still small. Second, local, non-mainstream scientific communication and international, mainstream scientific communication are interdependent and complementary.

483. de Bruin, Renger E., Robert R. Braam, and Henk F. Moed. 1991. "Bibliometric Lines in the Sand." *Nature* 349: 559-62.

Bibliometric analysis of citations and joint publications between Gulf State countries and U.S. Such cooperative activity tracks the political climate, including a reduction in U.S.-Iran linkages after the overthrow of the Shah, a general reduction of Iranian participation in Western science, the lasting effects of British and French colonial legacies, and the increasing importance of Saudi Arabian-American collaboration.

484. Desai, Ashok V. 1980. "The Origin and Direction of Industrial R&D in India." *Research Policy* 9: 74-96.

Finds a significant increase in Indian R&D expenditures, particularly in the corporate rather than in the government sector. A cross-sectional analysis of corporate R&D shows that larger companies are taking a "longer view" toward technological advances. Attributes the trends to import-control policies of two decades earlier, as well as a later recession.

485. Ebadi, Y. M. and D. A. Dilts. 1986. "The Relation Between Research and Development Project Performance and Technical Communication in a Developing Country: Afghanistan." *Management Science* 32: 822-30.

Examines effects of communication upon R&D project success. Data are from 49 government-funded R&D projects in Afghanistan during 1979. Results show that frequency of communication is positively related with performance for mid- and high-performing groups. Location of the project within the communications network (centrality) is negatively related to project performance within the low performance category.

486. Eto, H. 1991. "Science Revolution and Ortega Hypothesis in Developing Countries." *Scientometrics* 20: 283-96.

Relative weights of different scientific fields compared among countries, to indicate how much each values science. Differentiation among developing countries.

487. Fischer, William A. 1983. "The Structure and Organization of Chinese Industrial R&D Activities." *R&D Management* 13: 63-82.

Organizational and manpower problems result from limited economic development in a country. Language is more important than technical ability for the technical information offices. Low levels of education/training among technical and administrative staff mean that workforce quality is a more important constraint than organization of the system (especially those selected during the Cultural Revolution).

488. Fischer, William A. and Charles M. Farr. 1985. "Dimensions of Innovative Climate in Chinese R&D Units." ***R&D Management*** **15: 183-90.**

Uses Rickards' and Bessant's "Creativity Audit" in 141 R&D units (1982-3 data) and compares it with the original British data. Chinese units are more pessimistic than British units, less likely to say their organization operates to minimize individual risks. Their R&D programs are perceived to be less dynamic. These were relatively small units. General similarity in responses by Western and Chinese respondents.

489. Frame, J. Davidson and Mark P. Carpenter. 1979. "International Research Collaboration." ***Social Studies of Science*** **9: 481-97.**

International co-authorship patterns examined using the 1973 *Science Citation Index.* The more basic the field, the greater the proportion of co-authorships. The larger the national scientific enterprise, the smaller the proportion of co-authorships. International collaboration occurs along geographic lines. Table 4 shows cluster of countries according to international co-authorship patterns.

490. Frame, J. Davidson, Francis Narin and Mark Carpenter. 1977. "The Distribution of World Science." ***Social Studies of Science*** **7: 501-16.**

Uses corporate index of the *Science Citation Index* to survey international research activities. Publication data from 1973 for 117 countries. Mainstream scientific research is unequally distributed across nations, more so than land, population, and economic production. Ten largest producers account for 84% of the literature. Table 4 lists 15 major LDC science producers. India towers above other LDC producers by a factor of nine.

491. Frisbie, Parker, and Clifford Clark. 1979. "Technology in Evolutionary and Ecological Perspective: Theory and Measurement at the Societal Level." ***Social Forces*** **58: 591-613.**

Multiple factor index of technology regressed on economic growth, urbanization, political modernization, bureaucratization, and phase in demographic transition.

492. Frisbie, Parker, Lauren Krivo, Robert Kaufman, Clifford Clark, and David Myers. 1984. "A Measurement of Technological Change: An Ecological Perspective." ***Social Forces*** **62: 750-66.**

Finds that the process of technological change accelerated in the 1950-70 period in sample of 66 countries, but relative positions of countries remained the same.

493. Gaillard, Jacques. 1991. ***Scientists in the Third World.*** **Lexington: University Press of Kentucky.**

Survey of 489 scientists in 67 developing countries who received research subsidies from the International Foundation for Science. This volume is one of the best available demographic profiles of Third World scientists, though Asia is underrepresented and the respondents are not "representative" scientists from their home countries. Covers training, salaries, problem choice, institutional contexts, funding, equipment difficulties, communication practices, productivity, and publication practices--as well as national scientific communities in Costa Rica, Senegal, and Thailand. Research careers of Third World scientists are short because of rapid promotion to administrative positions, especially for those who spend more time abroad. Problem selection is influenced by international funding organizations.

494. Garfield, Eugene. 1983. "Mapping Science in the Third World." ***Science and Public Policy*** **10: 112-27.**

An examination of Third World research that stresses the unique traits of Indian science (e.g., physical sciences, biotechnology), and a high Indian success rate in achieving publication. Expectations for success by Third World scientists in the international research community, however dismal, are inflated by including India in the data.

495. Glyde, Henry R. and Virulh Sa-yakanit. 1985. "Institutional Links, An Example in Science and Technology." ***Higher Education in Europe*** **10: 51-9.**

Describes a linked cooperative program between the physics department at Chulalongkorn University in Thailand and the University of Ottawa, establishing R&D labs and a PhD program.

496. Goldstone, Leo. 1977. "Improving Social Statistics in

Developing Countries." *International Social Science Journal* 29: 756-74.

How statistics should be organized, including a section on characteristics of stats in developing countries.

497. Gorman, Lyn. 1981. "The Funding of Development Research." *World Development* 9: 465-84.

Statistical data on funding of OECD country research from 1973-79. Also looks at funding from government agencies and private groups. Emphasizes the importance of public funding.

498. Haas, E.B. 1980. "Technological Self-Reliance for Latin America: the OAS Contribution." *International Organizations* 34: 541-70.

Examines the OAS Regional Scientific and Technological Program, which supports R&D centers in Latin America and encourages division of labor among them. Finds that it failed to establish a regime because governments were unable to decide what they meant by "self-reliance," and thus could not choose a single norm that would orient activities with sufficient rigor.

499. Hill, Stephen. 1986. "From Light to Dark: Seeing Development Strategies through the Eyes of S&T Indicators." *Science and Public Policy* 13: 275-254.

Suggests that S&T indicators, although used as a basis for policy assessment in developed countries, must be translated carefully for application in developing countries. The use of indicators is now accepted for guiding development strategies. Gives examples of three levels of indicators for developing countries--indicators for planned national objectives (such as GNP per person as an indicator of distribution of wealth), indicators of sectoral status like competitiveness (such as local profitability versus profitability in equivalent international sectors), and indicators of receptivity to national S&T inputs like modernity (such as capital intensity per employee). Concludes by summarizing characteristics of indicators.

500. Howells, J. 1990. "The Internationalization of R&D and the Development of Global Research Networks." *Regional Studies* 24: 495-512.

Major trends in research and issues concerning the internationalization of R&D, with a focus on pharmaceuticals and chemical/energy sectors. Major aspects of globalization: flows of research inputs across national boundaries; supply and migration of personnel; links between R&D labs and other corporate functions; growth of interorganizational research collaboration. Two main models (focusing on supply and demand) in the literature on R&D and international production. Types of R&D unit and their structures in global and host-market firms are discussed, as well as factors leading to overseas R&D. Brief account of the internationalization of R&D in a UK pharmaceutical company from 1968-92.

501. Hsieh, H. Steve. 1989. "University Education and Research in Taiwan." Pp. 177-214 in *Scientific Development and Higher Education*, edited by Philip G. Altbach, et al. New York: Praeger.

Overview of the higher education system in Taiwan. Describes the scientific research system and its performing agencies. Statistics on national research and development, education, and R&D in Taiwan.

502. Huang, Y. 1986. "R&D in the People's Republic of China." *R&D Management* 16: 89-96.

Evolution of Chinese R&D structure and policy by author at Shanghai Polytech University's management department. China is seeking to convert R&D away from "catching up" to the world S&T system, which was the goal in the 1950s, and moving it in 1981 toward local needs and production. New competition between R&D institutes for industrial sponsorship. Old structure copied from Soviet Union in 1950s. Project selection was based on whether the research was up to world levels; no linkages to performance criteria. Developmental institutes are now encouraged to seek their own funding.

503. James, Dilmus D. 1990. "Science, Technology, and Development." Pp. 159-78 in *Progress Toward Development in Latin America: From Prebisch to Technological Autonomy*, edited by James L. Dietz and Dilmus D. James. Boulder: Lynne Rienner Publishers.

Addresses two questions: (1) How much freedom should institutes and investigators have in selecting research projects? (2) How much integration should S&T policy strive for? Then proffers policy suggestions. Includes remarks on institutional inefficiencies, critical minimum thresholds for research projects, and lack of relevance.

504. Jimenez, Jaime and Juan Escalante. 1990. "J.D. Bernal in a Latin American Perspective: Science for Development?" *Science Studies* 1: 59-66.

General background information on the development of science and technology in Mexico and the problems facing Mexican scientists today. Two sets of scientists: internationally oriented who publish in foreign languages and work on pure science, and "national" scientists whose research is applied and less visible.

505. Jimenez, J., M.A. Navarro and M.W. Rees. 1986. "Scientific Research Areas in Mexico: Growth Patterns in the Late Seventies." *Scientometrics* 9: 209-21.

A study of 10 Mexican research areas that dichotomizes research institutions as "primary" or "secondary" based upon quality. Concludes that secondary institutions face severe barriers to upgrading: high staff turnover, a national S&T system that does not facilitate improvements, few research personnel, poor distribution of resources. Suggests that upon creation institutions must be provided the resources necessary to be first rank, and that special long-range strategies will be necessary to upgrade existing secondary institutions.

506. Jimenez, J., P. Hunya, M. Bayona and A. Halasz. 1988. "The S&T Potential of Mexico and Hungary." *Scientometrics* 14: 17-41.

A comparative study of Hungarian and Mexican research institutions that finds the Hungarian research units are superior to Mexico's for a wealth of reasons: larger scientific infrastructure, older organized body of researchers, more practical orientations, stronger link with production sector, more financial support, greater international orientation, and more frequent publications.

507. Konrad, N. 1990. "Science, Technology, and Development Indicators for Third-World Countries: Possibilities for Analysis and Grouping." *Scientometrics* 19: 245-70.

Attempts to distinguish among countries according to: (1) a society's generativity (ability to generate a scientific and technological potential), (2) the potential itself, and (3) receptivity (the capacity to absorb or receive research results). A comparison of 30 developing countries shows: (1) the view that national development is tied to S&T potential is confirmed, (2) joint indicators offer better interpretations than comparisons of pairs of single indicators, (3) countries with comparable levels of the three capacities above will differ mostly in structure, and (4) the level of R&D potential is more closely related to the ability to absorb S&T results than to its resources for building up this potential.

508. Krauskopf, M., R. Pessot and R. Vicuna. 1986. "Science in Latin America: How Much and Along What Lines?" ***Scientometrics*** **10: 199-206.**

Reports a study of scientific output in Latin America and the Caribbean that evaluates the scientific "mainstream literature." Five nations--Brazil, Argentina, Mexico, Chile, Venezuela--produce 92% of Latin American output. Although most scientific research is found to be in the life sciences, the output in all fields appears insufficient "to assure a positive role of science for the best overall development of each individual society." Includes a citation analysis. The authors attribute some of this paucity to a lack of support for scientific progress.

509. Krishna, Venni Venkata. 1988. "Scientists in Laboratories: A Comparative Study on the Organisation of Science and Goal Orientations of Scientists in CSIRO (Australia) and CSIR (India) Institutions." ***Dissertation Abstracts International, A: The Humanities and Social Sciences*** **48: 2164-A.**

Scientists in two institutions are governed by distinct traditions. Research cultures are specific to the national and cultural context of the laboratory. The two organizations are in constant negotiation with the surrounding socio-economic and political environment. Research on goal orientations of scientists demonstrates that the meaning of research action is a product of influences routed through research cultures extant in the lab's historical context.

510. Lall, Sanjaya. 1979. "Transnationals and the Third World: The R&D Factor." ***Third World Quarterly*** **1: 112-8.**

Explores the possibility that transnational corporations can contribute to the strengthening of R&D capabilities in developing countries. Presents a speculative argument that, while the relocation of technological activity by MNCs to Third World countries is still relatively small, this relocation will probably grow relatively rapidly in the future. The economic forces that now make for comparative advantage in production are producing a shift in comparative advantage for R&D.

511. Lee, Chong-ouk. 1988. "The Role of the Government and R&D Infrastructure for Technology Development." ***Technological Forecasting and Social Change*** **33: 33-54.**

Director of the Korean Advanced Institute of Science and Technology examines rationale for government intervention. Presents the history of Korean industrialization in three stages. Institutional frameworks, legal systems, and manpower development are discussed in the context of Korean S&T policy for industrialization. Suggests strategies for the next 15 years.

512. Lee, Jinjoo, Sangjin Lee and Zong-tae Bae. 1986. "The Practice of R&D Management: An Empirical Study of Korean Firms." ***R&D Management*** **16: 297-308.**

Questionnaire study of 73 Korean R&D labs (those registered with the Korean Industrial Research Institute, 50% response rate) showing that management is still unstructured and flexible, although efforts are being made to enhance efficiency. Some contrasts with U.S. studies are included. No labs had fully adopted a matrix organization. From 1970 to 1984, the percentage of R&D expenditures by the private sector has increased from 12.6% to over two-thirds, while the percentage in public research labs has declined from 84% to 23%.

513. Lee, Sungho. 1989. "Higher Education and Research Environments in Korea." Pp. 31-81 in ***Scientific Development and Higher Education*****, edited by Philip G. Altbach, et al. New York: Praeger.**

Considers the contribution of academic science to national development in the Republic of Korea. Provides a general introduction to the educational system in Korea. Attempts to analyze the academic research environment and describes both institutions and fields. Investigates the national research system and how it takes shape in the academic system. Fifteen useful statistical tables on academic institutions and academic R&D in Korea.

514. Lenski, Gerhard and Patrick Nolan. 1984. "Trajectories of Development: A Test of Ecological Evolutionary Theory." *Social Forces* 63: 1-23.

Uses a sample of 77 countries to show that those with horticultural, preindustrial economies follow a different trajectory in the industrialization process than those with agrarian preindustrial economies.

515. Lewis, Robert. 1979. *Science and Industrialization in the USSR: Industrial R&D, 1917-1940*. London: Macmillan.

Manpower, money, and organization are the focus of this account of the USSR as the first state to plan research for state purposes. The Soviet system of R&D organization was an offshoot of an economic planning system that hindered innovation.

516. Lipsey, Robert E., Magnus Blomstrom and Irving B. Kravis. 1990. "R&D by Multinational Firms and Host Country Exports." Pp. 271-300 in *Science and Technology: Lessons for Development Policy*, edited by Robert E. Evenson and Gustav Ranis. Boulder & San Francisco: Westview Press.

Looks at the influence of multinational firms upon exports from developing countries, and the connection between R&D and multinational investment. Concludes that over the past 30 years multinational firms from the U.S., Sweden, and Japan have increased their shares of exports from developing countries. The generally held connection between foreign investment and R&D activity is supported--firms engaged in foreign investment were found to spend more on R&D relative to total sales and to have more employees involved in R&D than firms not engaged in foreign investment.

Firms investing in developing countries were found to be more R&D intensive than firms investing in developed countries. U.S. MNCs were found to perform less R&D in developing countries, relative to the size of their operations in those countries, than in developed countries and less in developed countries outside the U.S. than in the U.S. Differences among MNC home countries in domestic R&D intensity do not seem to be related to composition of trade. The proportion of high-tech industries in country exports is higher when U.S. affiliates in the country engage in intensive R&D.

517. Lomnitz, Larissa A., Martha W. Rees and Leon Cameo. 1987. "Publication and Referencing Patterns in a Mexican Research Institute." ***Social Studies of Science*** **17: 115-33.**

Traces the 20-year development of a Mexican research agency that was successful on the international level, and analyzes that development.

518. Long, Frank and Gillian Pollard. 1984. "An Examination of Patent Statistics in Guyana, 1903-1980." ***Studies in Comparative International Development*** **19(3): 3-14.**

Finds that purportedly restrictive patent laws in Guyana actually benefit foreigners. Uses statistics from the Deeds Register of the Patents Registry of Guyana, from 1903 to 1980, to show that foreign patents have been on a relative rise over the years, while local patents have shown a secular decline. Concludes that patent policy in this country has not been an incentive to indigenous technology and that the increase in foreign patents has not led to structural changes in the economy. Suggests radical reform in the patent system.

519. Luukkonen, Terttu, Olle Persson, and Gunnar Sivertsen. 1992. "Understanding Patterns of International Scientific Collaboration." ***Science, Technology, and Human Values*** **17: 101-26.**

Interpretation of macro data on international co-authorships. Cognitive, social, historical, geopolitical, and economic factors as determinants of country differences in the rates of co-authorship and networks of collaboration. Methodological approaches to propensities to collaborate, independent of country size.

520. Moravcsik, Michael J. 1982. "The Effectiveness of Research in Developing Countries." ***Social Studies of Science*** **12: 144-50.**

This is a response to Norman Clark's article in volume 10 and is followed by a rejoinder from Clark. Challenges Clark for a number of assumptions: (1) that the sole legitimate objective of developing countries is to improve economic status; (2) that a country will have "national objectives"; (3) that a nation has an appropriate level of trained people to enact recommended changes; (4) that research institutions often can handle recommended changes.

521. Moravcsik, M.J. 1985. "Applied Scientometrics: An Assessment Methodology for Developing Countries." ***Scientometrics*** **7: 165-76.**

Describes a project, undertaken by the Centre for Science and Technology for Development of the United Nations, that had the dual aim of formulating a practical assessment system for S&T in developing countries and formulating a research program aimed at making improvements on systems of assessment. The author describes the first step in the assessment process as making a map of the part of the system to be assessed. After this, types of indicators are listed. The author suggests that the status of these indicators is weak, especially with regard to developing countries. He proposes that a small number of pilot projects be undertaken to test the ideas in the discussion and to experiment with new kinds of indicators.

522. Morita-Lou, Hiroko (ed.). 1985. ***Science and Technology Indicators for Development.*** **Boulder, CO: Westview Press.**

Important volume suggesting that the indicators used to measure the impact of science and technology in developed countries often lead to inaccurate conclusions when applied to LDCs. Discusses strengths and limitations of current indicators.

523. Nagpaul, P.S. and S.P. Gupta. 1989. "Effect of Professional Competence, Managerial Role, and Status of Group Leaders on R&D Performance." ***Scientometrics*** **17: 301-32.**

An analysis based upon data collected from 1,460 research units in six countries. Studies the relationship between measures of leadership qualities possessed by research unit leaders and measures of performance, such as image of quality, and indexes of scientific, user-oriented, and administrative effectiveness. Professional and managerial competition found important for image and effectiveness, especially professionalism for effectiveness. Finding is general across the countries studied.

524. Nagpaul, P.S. and V.S.R. Krishnaiah. 1988. "Dimensions of Research Planning: Comparative Study of Research Units in Six Countries." ***Scientometrics*** **14: 383-410.**

A study of six nations' research units: Argentina, Egypt, India, Republic of Korea, Poland, and the USSR. Three items clearly predict performance: (1) specificity of research goals, (2) conceptual challenge of the research program, and (3) the quantity and quality of the group's external linkages with scientific peers and potential users of their research results.

525. Nussenzveig, H.M. 1992. "Research Funding in Brazil: A Case History." *Technology In Society* 14: 137-50.

Analyzes the history of research funding in Brazil. Also includes a discussion of modern strategies to increase peer participation in R&D policy planning, stabilize research groups and encourage informed public opinion about S&T matters.

526. Oral, Muhittin, Nukhet Yetis and Riza K. Uygur. 1981. "Participatory Planning of Industrial R&D Activities." *Technological Forecasting and Social Change* 19: 265-77.

Describes methodology for participatory planning of R&D activities in Turkey's iron and steel industry. This industry consisted of three integrated iron and steelworks complexes. During the 1980s hundreds of managers, engineers, planners, and experts were brought together by the State Planning organization of Turkey to prepare their own planning for R&D activities. Study team provided information, instructions, coordination, motivation, and facilities. Four phases of the planning are identified and described: (1) studying the industry, (2) identifying needs and problems, (3) formulating R&D projects, and (4) evaluating, selecting, and managing R&D projects.

527. Rabkin, Yakov M., Thomas O. Eisemon, Jean-Jacques Lafitte-Houssat and Eva McLean Rathgeber. 1979. "Citation Visibility of Africa's Science." *Social Studies of Science* 9: 499-506.

A study of citations for papers from two African universities in the fields of botany, zoology, mathematics and physics. The data are compared to that for two other peripheral universities. Botany and zoology papers were found to be more locally cited. But in general, the authors found consistently high visibility for the scientific work, both in Britain and intranationally, leading them to conclude that the scientific work of developing nations receives "significant visibility" in scientific communications.

528. Rajeswari, A.R. 1983. "A Quantitative Analysis of Indian Science and Technology Manpower Employment and Economic Development." *Scientometrics* 5: 343-59.

Looks at the relation between S&T employment and development of the country.

529. Reynolds, John and Lauren Krivo. 1992. "Change in Societal Technology, 1950-1985." Paper delivered to the American Sociological Association.

Extends Frisbie's multidimensional technology measurement from 1950-1985 (51 nations, underrepresenting less-developed countries) and from 1970-1985 (96 nations). Main dimensions: energy, agriculture, manufacturing, transportation, and communication. Stronger relationship between technological change and social well-being shows that it is a better measure than energy consumption of societal technology (and most technologically advanced nations had decreasing rates of energy consumption in the 1980s).

Total change in the measure of technology was decomposed into change in average level, dispersion, and positional change (ordering of countries). Results for the shorter time period (more representative sample) show that positional change is more important than either dispersion or mean change. Increases in dispersion (technological distance between nations) was of little importance after 1970.

530. Roche, M. and Y. Freites. 1992. "The Rise and Twilight of the Venezuelan Scientific Community." *Scientometrics* 23: 267-90.

Offers a statistical description of the Venezuelan scientific community in Caracas, during the period 1976-78. Based on data collected in individual interviews with a stratified random sample of 473 scientists from the three main research organizations in metropolitan Caracas. Observes that since 1982, as a result of inflation and devaluation, facilities, pay, and work satisfaction have declined, leading to a migration of scientists abroad or to industry.

531. Ronstadt, R. 1977. *Research and Development Abroad by U.S. Multinationals*. New York: Praeger.

Analyzes the foreign R&D activities of seven multinational corporations: Exxon Corporation's energy businesses, Exxon Chemical

Company, International Business Machines, Chemicals and Plastics Group of Union Carbide Corporation, CPC International, Otis Elevator Company, and Corning Glass Works. Findings are based on interviews conducted by the author in 1974 with about 60 executives in these organizations.

Part I contains descriptions of foreign R&D activities. Part II is an aggregate analysis of the establishment and evolution of foreign R&D activities by these seven MNCs. Among the conclusions: R&D activities followed a definite trend, often developing into indigenous technology units in order to keep their best R&D professionals by offering the opportunity for more challenging projects. Ends with a number of administrative implications of the R&D process abroad.

532. Sagasti, Francisco R. 1989. "Crisis and Challenge: Science and Technology in the Future of Latin America." *Futures* 21: 161-8.

Broad discussion of problems facing Latin America. Projections based on past trends indicate that in the year 2000 the region will spend about .55% of GDP on R&D, slightly above current level. If R&D expenditures were to grow at 10% per year (unlikely), then R&D spending in 2000 would be about .72% of GDP, a figure comparable to small European countries at beginning of 1980s. These upper-limit levels of spending don't lead one to expect any great leap forward in regional scientific and technological capacity.

533. Sakai, Fuminori. 1986. "Collaborative Efforts in Science Between Japan and Other Countries: Past, Present and Future." *Perspectives in Biology and Medicine* 29: S57-S65.

Includes data and analysis of Japan's scientific collaborations with nations across the globe, including developing countries in South East Asia.

534. Schott, Thomas. 1988. "International Influence in Science: Beyond Center and Periphery." *Social Science Research* 17: 219-38.

Old models of the global network of scientific influence use a simple center-periphery conception. Using reference patterns from the *Science Citation Index*, a network model generates six geopolitical regions of structurally equivalent communities (Latin America, Eastern Europe, Central Europe, Scandinavian, Australian, Anglo-Saxon) explained by linguistic affinities, collegial and educational ties, and political barriers between countries.

535. Schott, Thomas. 1992. "The World Scientific Community: Globality and Globalisation." ***Minerva*** **30.**

Examination of the integration of scientific communities across national boundaries. Citation data show a center/periphery structure, with the liberal democracies most influential but the communist countries a somewhat separate network. Publications data show the share of LDCs in world science. Finally, co-authorship data, place of publication, and percentage of citations outside the country show relative orientations to foreign audiences. Scientists in all countries have increased the extent of their collaboration with foreign scientists from 1973-1986.

536. Schubert, A. et al. 1990. "World Flash on Basic Research. Scientometric Datafiles. Supplementary Indicators on 96 Countries 1981-85." ***Scientometrics*** **18: 173-8.**

A quantitative summary of journal publication activity by scientists in dozens of nations. Although the compilation is of specific interest, this journal also often has specific information on the quantity and quality of research being produced on a comparative national basis.

537. Sen, B.K. and V.V. Lakshmi. 1992. "Indian Periodicals in the ***Science Citation Index*****."** ***Scientometrics*** **23: 291-318.**

Finds that coverage of Indian S&T periodicals in the *SCI* during the period 1975-88 is poor. Many of the journals do not fulfill criteria for inclusion in the *SCI*. Identifies about 500 periodicals which are covered by at least one major world indexing or abstracting service. The total number of these periodicals is likely to be over 600. These probably can be made worthy of coverage by the *SCI* with slight improvements.

538. Sen, Falguni. 1988. "The Dilemma of Managing R&D in India." Pp. 279-291 in ***Science, Technology, and Development*****, edited by Atul Wad. Boulder: Westview Press.**

Gives a general description of the nature of R&D management in a developing country, characteristics of R&D specific to India (such as India's large scientific research base), and expectations from R&D in different sectors. Eight issues are central to R&D management in

India: (1) the appropriateness of a bureaucracy in managing an activity like R&D, (2) the tendency of laboratories to become dominated by their directors, (3) deciding on optimal resource allocations for different research projects, (4) better coordination of research activities conducted by different agencies (capital scarcity makes it necessary to avoid duplication), (5) the frustration of scientists, (6) the likelihood of implementation of scientists' ideas, which is increased by scientists' awareness of the real needs of the sectors they are trying to influence, (7) a scarcity of resources that is not conducive to individual creativity, and (8) finding ways of effectively transferring technology. Resolving these issues depends partly on clearly articulating the expectations from R&D in each sector, and partly from achieving a more decentralized, flexible organization for R&D management.

539. Sharif, M. Nawaz (ed.). 1987. "Technological Capabilities Assessment in Developing Countries." *Technological Forecasting and Social Change* 32(1): 1-109.

A special issue devoted to assessing the technological capabilities of developing countries. The articles in this issue are the work of a project entitled "Technology Atlas," which was initiated as a result of a special study prepared by Sharif for the 40th annual session of the United Nations Economic and Social Commission for the Asian Pacific (UN-ESCAP), held in Tokyo in April, 1984. These articles, authored by the Technology Atlas Team, cover methodologies being developed by the Asian and Pacific Centre for Transfer of Technology (APCTT) of ESCAP.

The first paper provides the justification and need for national and international initiatives for technology-based development. The second paper maintains that technology is a combination of "Technoware, Humanware, Infoware, and Orgaware". The fourth and fifth papers show how to use the conceptual framework proposed by the Atlas team to assess gaps and levels of development in specific technologies. The final paper compares the national technology climate in Thailand and Sri Lanka.

540. Singh, Jasbir Sarjit. 1989. "Scientific Personnel, Research Environment, and Higher Education in Malaysia." Pp. 85-136 in *Scientific Development and Higher Education: The Case of Newly Industrializing Nations*, edited by Philip G. Altbach, et al. New

York: Praeger.

Looks at the role of universities in the creation and development of scientific personnel, and examines the participation of universities in Malaysian R&D activities. Includes discussions of the scientific research system, R&D manpower and expenditure, higher education and high-level scientific manpower, case studies of two Malaysian universities, the research environment, and research productivity. Concludes that Malaysia's educational system has created sufficient high-level scientific manpower with credentials, but that doubts remain regarding the adequacy of this manpower to lead industrial development through increased R&D. Seventeen statistical tables on the Malaysian educational system and R&D institutes.

541. Singh, Prithpal and V.S.R. Krishnaiah. 1989. "Analysis of Work Climate Perceptions and Performance on R&D Units." ***Scientometrics*** **17: 333-52.**

Reports findings from a study on perceptions of work climate, and patterns of relationship between work climate dimensions and performance of research and development units in Argentina, Egypt, Republic of Korea, Poland, and USSR. These are examined by types of institutions: academic, government research, and industrial research. Work climate is affected more by socio-cultural factors than by institutional settings and is important for the effectiveness of the R&D unit.

542. Spagnolo, F. 1990. "Brazilian Scientists' Publications and Mainstream Science: Some Policy Implications." ***Scientometrics*** **18: 205-18.**

Argues that, despite a common view that *Science Citation Index* underrepresents scientific output from peripheral and non-English-speaking countries, academic Brazilian scientists in chemical and electrical engineering tend to publish in "good" international journals covered in the *SCI*. However, they earn a low rate of citation. The author questions the policy of encouraging Brazilian scientists to publish in foreign journals.

543. Standke, Klaus Heinrich. 1987. "Science and Technology: The Situation in Africa." ***Development: Seeds of Change*** **2&3: 84-7.**

In the allocation of funds by UNDC to the 36 least developed countries (69% are African), only 1.6% of funds to went to science and technology between 1970 and 1980. Eleven reasons why science and technology have failed to develop. Recommends regional rather than national projects and networks, close relations with "paired" institutions in developed countries, and non-governmental organizations.

544. Stolte-Heiskanen, Veronica. 1987. "Comparative Perspectives on Research Dynamics and Performance: A View from the Periphery." *R&D Management* 17: 253-62.

Poses major research questions derived from the hypothesis that conditions on the periphery render problematic the determinants of productivity used for the center. Small, industrialized and developed European countries also would be considered peripheral in her conceptualization. Considers research group and institutional levels of analysis.

545. Technology Atlas Team. 1987. "Assessment of Technology Climate in Two Countries." *Technological Forecasting and Social Change* 32: 85-109.

Attempts to show that a preliminary database for technological climate evaluation can be developed using information already available in developing countries. This evaluation should be carried out under six headings: (1) level of development as measured by classical indicators, (2) evaluation of S&T personnel stock, (3) evaluation of R&D activities, (4) evaluation of technology in industry, (5) evaluation of academic S&T, and (6) measures for promoting and popularizing S&T.

The suggested contents of such a database are illustrated with data from Thailand and Sri Lanka during the early to middle 1980s. Data from Japan are used in a few of the tables for the sake of comparison. The 29 tables contain statistics on GDP, demographics, health indicators, student enrollment, mass communication, S&T personnel, expenditure on R&D, export to import ratios, patents and trademarks, degree-granting institutions, S&T papers published, and level of commitment to S&T. The article is highly recommended, both for the general study of S&T development indicators and for the study of S&T development in Thailand and Sri Lanka.

546. Technology Atlas Team. 1987. "Measurement of Level and Gap of Technological Development." *Technological Forecasting and*

Social Change **32: 49-68.**

Outlines a procedure with which to measure levels of development, particularly the "gap" between two industries, countries, or firms. Cooperation is most profitable with moderate gaps, rather than large or small ones. Gaps are a function of differences in Technoware, Humanware, Orgaware, and Infoware. Examples from Japan, India, and Korea in iron/steel, rice, electronics, and computer technology.

547. Teitel, Simon. 1987. "Science and Technology Indicators, Country Size and Economic Development: An International Comparison." ***World Development*** **15: 1225-35.**

Using UNESCO 1980 and 1982 data, regresses stock of experts, number of scientists in R&D, and R&D expenditures on population size and per capita income. Models need nonlinear effects, so average coefficients for population size and per capita income should not be used, and the practice of using them for all countries should be discontinued. Separate models needed for industrialized and less-developed countries.

548. Velho, Lea and John Krige. 1984. "Publication and Citation Practices of Brazilian Agricultural Scientists." ***Social Studies of Science*** **14: 45-62.**

Study of publications by Brazilian scientists reveals a fragmented scientific community in that nation, with poor links to colleagues in different institutions and lagging research work.

549. Vessuri, Hebe M.C. 1986. "Universities, Scientific Research, and the National Interest in Latin America." ***Minerva*** **24: 1-36.**

Excellent, brief history of universities in Latin America, their conservatism, and their shortcomings with respect to teaching and research. Difficulty in translating patterns of research into a local context and creating a research ethos at either undergraduate or graduate levels. Conflicts exist between science policy and education policy of a country. Article provides data on students by country, as well as research expenditures by sector for Brazil, Mexico, and Chile. Discusses national scientific communities and their relations with government and society, along with the "mass universities" of the recent past. The "brain drain" is described briefly, as are connections between academic research and industry, and future prospects for science and the university.

550. Vessuri, Hebe M.C. 1987. "The Social Study of Science in Latin America." ***Social Studies of Science*** **17: 518-54.**

Discusses the growth and change in social studies of science in Latin America, and includes substantial data about the changing role of science and technology for innovation in that region. Includes stats on R&D and state policy planning. Most current work is on the subjects of formal institutions of science, scientific communities, social history of disciplines, science policy, and technological innovation.

551. Visvanathan, Shiv. 1985. ***Organizing for Science: The Making of an Industrial Research Laboratory.*** **Delhi: Oxford University Press.**

History focusing on the National Physics Laboratory of India, with excellent treatment of the conflicts in orientation between basic and applied research. Divides the rise of industrial research in India into three phases: (1) development of field organizations to perform surveys in the 18th century, (2) establishment of universities with a research function, and finally, during the nationalist movement, (3) the need to link science and technology through the establishment of agricultural research and the industrial research laboratory. Covers historical debates over the priority of science or technology, the use of the Soviet model, the differences between science for industry and science for agriculture, the debates of the Science and Culture Group. A vision of industrial research is linked to the vision of a planned industrial society on the Soviet model. The decision to create a network of research laboratories (CSIR) was based on the realization that the industrial revolution in the West was not based on any systematic use of science, but a belief that investment in science would pay off. Protected scientists under Nehru fostered this myth. Industrial research in the lab arose as the "scientization of technology," and was followed by the "industrialization of science." Four case studies showing technology transfer to industry. Final chapter considers role of the lab in trying to prevent the destruction of a primary product's market (mica), and in creating a market for mass-produced radio in a nation of villagers (both failures). Conclusion discusses industrial research laboratories as technology-producing factories.

552. Wionczek, Miguel S. 1983. "Research and Development in Pharmaceuticals: Mexico." *World Development* 11: 243-50.

Finds links between low local R&D and the importation of foreign technology. Found U.S. firms in the area spent little on R&D, instead focusing on the importation and advertising of technologies.

553. Yuthavong, Yongyuth. 1986. "Bibliometric Indicators of Scientific Activities in Thailand." *Scientometrics* 9: 139-43.

Examines the scientific output of Thailand by looking at the correlation between international publications covered by *Science Citation Index* with works published in the *Journal of the Science Society of Thailand*, and by the correlation of *SCI* publications with the annual symposium of the Science Society of Thailand. The correlation between number of articles from Thailand's 13 leading institutions covered by the *SCI* and works published in the *Journal of the Science Society* is .92, while that between *SCI* articles of Thai institutions and abstracts of the annual symposium is .73. (The higher correlation of the *Journal* with the *SCI* is explained by the fact that the *Journal* is in English, that it has more foreign contributors, and by the refereeing process.) These indicators reflect Thai scientific productivity. However, they do not indicate the extent of the impact and utilization of Thai scientific production.

Part III -- Non-Economic Causes of Stagnant Third World Development:
Culture, Politics and Institutions

Economic conditions in less-developed countries (LDCs) go a long way toward explaining the inability of LDCs to catch up with the West in scientific and technological development. Countries with abundant economic resources are better able to support the educational institutions, research laboratories and communications facilities required for a strong scientific community. Those nations with a strong technological base have the necessary foundation for further growth in technological capacity. This connection is obvious enough that many early scholars turned to these factors as the source of Third World economic problems, a materialist logic that is still highly influential.

Nonetheless, scholars increasingly have progressed beyond mechanical explanations to consider other factors that might explain LDC problems with scientific and technological growth (Crawford 887). Both the success of the Newly Industrialized Countries (NICs) and the repeated failure of other developing nations point to the inadequacy of purely economic discussions of Third World difficulties. Materialism might tell us why the North continues to dominate the South in science and technology, but it certainly fails to account for much of the variation among the nations of the periphery. To explain this diversity, we must bring in other variables beyond the poverty that most nations of the South share.

Of course, this observation is nothing new; some of the earliest modernization theorists pointed to socio-psychological factors as obstacles to what was seen as an otherwise linear path to development. Nevertheless, in the science and technology (S&T) field the big break from determinism showed up in the 1970s, among leftist scholars refining dependency theory. The World Systems approach, which emerged around that time, took into account both cultural and international political factors that could represent influences on a nation's S&T development. This insight would catch on as the many approaches to Third World poverty converged later in the decade (So 1990). Students of Third World scientific and technological stagnation also turned increased attention to political economy and institutional legacies within the LDCs themselves to understand varying performances.

The remainder of this essay will attempt to give a brief, functional sketch of the work contained in the bibliographic sections of this book, pulling together much of the best recent academic work on the interaction between non-economic factors and S&T development. The discussion is divided into three parts: Culture, Politics, and Institutions.

Culture

A nation's scientific and technological development does not take place in a vacuum. Human beings must staff the research laboratories, make the technology transfer decisions, and choose to accept or reject potential innovations. These individuals bring to their tasks varying technical capacities, levels of adaptability, cultural assumptions about technology, and habitual methods of decision making. When these traits are shared on a societal level, they may hamper or promote a nation's S&T development.

In general, cultural variables are invoked to explain why some nations of the South have failed to develop a scientific and technological capacity equal to that of neighboring Third World nations or of Western industrial countries. One set of such cultural factors includes: (1) A "pre-industrial mentality" among the citizenry (e.g., Ahmad 954). (2) An elite ideology or religious belief that influences technological choice and sometimes militates against modernization (Adler 555, 556; Ahmad 2). (3) A language ill-equipped to handle scientific concepts (Abdallah 952). (4) Inadequate human resources.

Even the most sympathetic discussion of these factors tends to "blame the victim" by emphasizing traits within an LDC culture that are problematic. Third World nations do not engage in scientific and technological development because something within their society stands in the way. On the other hand, these scholars are among the most optimistic about the long-term prospects for LDCs. They tend to see little about Western science itself that is incompatible with the

Third World, once a nation's internal obstacles are overcome or technical knowledge is adapted to fit into the recipient culture.

In contrast to the optimistic cultural assessment outlined above, two other views of the relationship between S&T and Third World culture tend to dominate the social science literature (Urevbu 977). One portrays Western S&T as totally incompatible with the cultures of the periphery. The other, a sort of compromise view, suggests that S&T transferred from the West must be adapted radically before it can thrive in the Southern Hemisphere. The two views have one element in common, however--they blame Western science for the inability or unwillingness of LDCs to "develop." Western technology and scientific education are seen to contain cultural assumptions alien and perhaps detrimental to the Third World; they embody choices made in the West over the trade-off between "development" and environmental damage or social disruption that Third World societies may not accept.

Barriers Within Indigenous Cultures

To an extent, all of the cultural factors described as barriers to S&T development qualify as products of a "pre-industrial mentality," since they are traits of indigenous societies that prove resistant to Western S&T. Scholars attempting to find symptoms of such a mind-set look for resistance to technological innovation or the scientific method. Sharafuddin (788), for example, argues that lack of acquaintance with scientific concepts is a serious handicap to Third World development. If S&T is to be popularized in the periphery, local culture and local languages must be enlisted. Gomezgil (921) applies this advice in particular when discussing the acculturation of Third World youth to Western science; the image of the scientist taught to them should be modified by "local characteristics." Rao and DuBow (832) suggest that, while scholars and policy makers realize the necessity for technology geared to Third World needs, they still neglect the importance of getting the target group for innovative technologies to overcome their suspicions and accept change. Processes of innovation within a society or its institutions can influence the speed of development (Fransman 374), as can the way a culture mediates its citizens' perceptions of the gravity of social change (Yuchtman-Yaar & Gottlieb 843). Finally, the economic benefits of development seldom do much to break down traditional resistance to change, since fertility rates eat up much of the payoff for innovation and result from cultural traditions most difficult to change without providing economic incentives (Hoshino 816).

While the above discussion stresses popular conceptions of S&T, another looks at elite conceptions, as expressed in religious doctrine or political ideology. Elite ideologies are blamed for stagnation in India (Ahmad 2) and the Arab states (Dorozynski 729). The role of religion is most commonly discussed with reference to China. Confucian traditions of deference are seen as highly resistant to the scientific method (Baum 564), even when the Chinese government pushes plans for "modernization" (Baum 565). These traditions are seen to dominate science and education in China even as Western models of school organization move in (Hayhoe 743). This suspicion of technology and experts carried over to Maoist political ideology, the most common example of which is the Cultural Revolution (Blecher & White 575; but see Chai 579). The Chinese stress on autonomy is seen to suppress coherent central planning (Blanpied 574). The state has stifled scientific development (Qian 655) and politicized education (Shirk 789). On the other hand, scholars have begun to notice changing attitudes to S&T in China, and not all are pessimistic about future prospects (Broaded 578, Conroy 725, Sigurdson 681).

Third World nations may suffer linguistic handicaps to development as well. Much scientific literature, for example, is published in English; the language is almost a requirement for access to scientific knowledge (Grabe & Kaplan 740). Eisemon and Rabkin (735) provide evidence that a multilingual society on average will have more difficulty with S&T development than a monolingual one, a relevant finding since LDCs commonly have many more languages than Western countries. Scholars have documented particular examples of difficulties translating scientific concepts into indigenous languages. Hewson and Daryl (746) describe how an African tribe utilizes cultural metaphors for heat that are resistant to Western, scientific conceptions of it. Watson (949) describes how the Yoruba tribe in Sub-Saharan Africa operates with different linguistic constructs for numbers.

Finally, cultural traditions combine with poor economic conditions to promote a neglect of human resources. The problems of inadequate human capital impact almost every level of S&T development. Alvarez (845) documents how the physiological problems associated with malnutrition pose tangible handicaps to education. A cultural emphasis on resource-oriented industry exacerbates the inequalities and unemployment produced by technological innovation

(Roemer 336). Jaireth (817) indicates, for example, that reliance on small cultivators in India's Punjab assures the dominance of inefficient agricultural production methods. In general, Third World nations lack the expertise needed to do quality research or make crucial technological choices (Banta 846, Haule 891, Hill 171, Lall 394). Worst of all, an LDC lag in human resources will pose an even greater threat as the service sector rises in importance in the global economy (Heitzman 892).

Few scholars dispute that low human resources or high language complexity can hinder a nation's ability to develop its S&T capability, although Lall (394) indicates that LDCs can succeed in industrial markets when intensive research is not required. On the other hand, the monolithic imposition posed by mass or elite ideologies has not been universally accepted. Some evidence points to the essential flexibility of people's attitudes toward S&T. Carroll (1984) finds no essential incompatibility between Third World religions and modern science, for example. Pattnaik (933) makes the same argument as it applies to Indian religions. Far from blaming elite ideology for low development, Moravcsik (638) attributes elite desire to match the S&T success of Western societies for many LDC development programs. Finally, a study of Nigerian faculty members--the educational elite--and their attitudes toward media technology suggests willingness to adopt innovative products and processes (Ajibero 711).

Barriers Within Western Science and Technology

Innumerable scholars have found a "Western bias" in science and technology (Buck 917, Choudhuri 721, Dugan 731), although exactly what this bias entails is seldom explicated. Most who make this argument disagree only over whether the cultural baggage can be expunged, with some envisioning a new form of science--for example Sardar's (937) call for a "return to Islamic science"--and others seeing the very scientific method as antithetical to Third World needs. Sardar (674, 976) has perhaps been most vocal about the threat posed by Western S&T to the "Muslim view of development" and the need for competing approaches to scientific progress from the periphery. Ziadat (951) outlines some of the difficulties Arab intellectuals faced in dealing with Darwinism and other Western scientific concepts. Jamison and Moock (753) argue that Western agricultural education actually may not lead to greater agricultural efficiency than traditional methods.

Handberg (742) claims that so-called Western models of science even pose problems in non-elite Western institutions. Finally, Gareau (738) posits that Third World scientists already have rejected many Western scientific paradigms quite successfully.

However, these descriptions of Western, capitalist bias in S&T have not garnered anything near universal acceptance among academics either in the West or the South. Childers claimed as early as 1979 that people were progressing beyond the assumption the West must lead in S&T development (958). Kettani (968) points out that inductive reasoning is not a Western invention as its critics apparently presume; indeed it might be an Arabic one. Zahlan (709) adds that Arabs are more committed to science than is commonly perceived. Needham (929) shows that the Chinese led the world in technical innovation until the 16th century. Bass (3) finds high scientific potential during his travels across the African continent. Simmons and Alexander (46) conclude that the determinants of academic success are pretty much the same in the Third World as they are in the West, which seems at odds with claims Western science is incompatible with LDC cultures. Finally, Teitel (105) challenges the assumption that Third World nations passively accept Western technologies--like some sort of cultural Trojan Horse--blind to the cultural biases lurking within them. He argues that decision makers actively choose among technologies and adapt them to local needs; many of the mistakes in technology transfer are the result not of cultural inappropriateness but of the failure to meet normal market tests.

Less controversial is the claim that technologies established in the West carry with them potential for severe social disruptions that many in the Third World will not be willing to accept (Nichols 828, Richter 833). These views are less controversial only because they are not inherently ideological; one can accept that technology carries severe costs and still argue over the relative value of costs and benefits. Maier (898), for example, observes the important social impact of information technologies among young Chinese students, a prophetic observation in light of disruptive pro-democracy student protests in the late 1980s. Malik (822), observing experiences in North India, warns that the cultural change accompanying technological innovation can cause alienation among potential users. Technological development makes armed revolts more likely and increases probability of success (McDonald & Tamrowski 399). Tobey (870) points out that Third

World nations are least capable of heading off the environmental damage that accompanies Western technology. Schiller (908) offers a classic argument that information technology allows the core nations a subtle means of cultural domination, replacing more costly policies of outright colonialism. The most worrisome disruptions resulting from technological innovation, however, are frictional unemployment and economic inequality. Any capital-intensive production method that increases efficiency is likely to cause at best short-term disruptions in the labor force, at worst self-perpetuating cycles of climbing inequality. Nilsen (903) finds such results, for example, to stem from LDC computerization. As Rahman (973) argues, even if the economic suffering brought on by innovation is temporary, it can drive a wedge between S&T and the society expected to accept it.

Politics

Although scholars were quick to pick up on the cultural barriers to technological diffusion, the integration of political economy in the study of S&T was more gradual (Barnett 355). As late as 1983, Henriquez felt the need to call for increased awareness of political actors in studying S&T issues. In the decade since his plea, numerous academics have risen to the challenge, churning out a generous body of literature studying political influences on scientific capacity and technology choice (Kenney 758).

Within the political subfield of S&T literature, few scholars dispute the significant influence governments can have on development through implementation of policies regulating trade, research or engineering. The disagreement instead centers on the normative question of whether the state *should* attempt to direct S&T policy. While some scholars tend to castigate Third World governments for not being active enough in the promotion of science, or at least for failing to enact policies optimal for development, another school faithfully delineates large bodies of evidence showing that government interference in S&T is the cause of many sub-optimal development outcomes. In effect, the debate breaks down into the classic division between those who see a significant role for government in addressing market failures and those who consider government likely to distort normally healthy market processes and hinder optimal technological choice.

However, the national government is not the only political

actor with a significant influence on S&T. Other scholars have turned their gaze outside of underdeveloped Third World nations, searching for international influences on either national governments or the process of development itself. While some tend to be critical of international influences on Third World development of science and technology, a few find examples of international organizations or foreign governments actively and successfully promoting LDC improvements.

Governmental Influence

Books and articles calling on Third World governments to pass this or that science policy are not difficult to find (Edquist & Jacobsson 806, Hoffman & Rush 894, Hurst 384, Mtewa 640). Governments are seen as capable of formulating policies that will successfully aid in the S&T development process (Altbach 877, Berlinguet 567, Macioti 634), and in limited conditions may be successful in operating exporting enterprises themselves (Ramamurti 657). Some scholars even provide specific examples of positive government influence on a nation's technological capacity (Katz & Marwah 857). Hobday (614) credits Brazil's government with facilitating that country's success in telecommunications and digital technology. Westman (915) adds that the Brazilian government successfully helped the country retain control of its microcomputer industry despite intervention by U.S. corporations. Wise (707), using Brazil and Argentina as case studies, argues that space technology can influence development positively if embedded in an optimal policy environment. Toure (791) describes a government propaganda campaign conducted on the Ivory Coast that effectively popularized science at the grassroots level. Drawing on the Korean example, Choi (584, 585) sees an important government role in S&T development -- linking science and industry, promoting education, and investing in long-term, high-risk research projects.

Others criticize political elites for not promoting particular S&T policies, blaming a lack of scientific and technological progress on too little government interference in a certain area rather than too much. Wionczek (706), for example, criticizes the Mexican government for operating on an uncoordinated S&T policy. Bowonder (848) suggests India has inadequate policies regulating toxic materials. Ramanathan (658), as part of a discussion of the Sri Lankan case, calls for government intervention to expand from mere distribution of expenditures or personnel to the crucial task of distributing technical information. Jairath (752), outlining in detail how political concerns have undermined India's personnel and spending policies, nevertheless sees a role for government in pushing "vast social change." Economists Roe and Shane (335) find that a government can promote high-tech industrial concerns by providing marketing assistance, arguing that such aid can do more than lowering the industry's factor costs through subsidies. Randolph and Koppel (659) argue that some governments should pay greater attention to reasonable assessment of technologies, rather than forcing a rush for rapid technological adoption. Finally, Rushing and Brown (662) have published a collection of essays with the general theme that poor intellectual property protections hinder Third World research and development.

Yet all of these clarion calls for government intervention stand in bold contrast to a huge body of evidence that political interference distorts industrial markets (Hill 611, Marton & Singh 326). Turnbull (1989) discusses this dynamic for Papua New Guinea's vaccine development, and Warhurst (461) for how it applies to metals biotechnology. Policies tend to respond, not surprisingly, to political needs, but are insulated from market demands or local community requirements (Melody 823, Moore 971, Ranis 660; also Whiston 704 on education policy). Government programs are biased toward showy capital-intensive projects rather than the most efficient choices (Perkins 652). Examples include excessive Israeli investment (Bregman 577), Indira Gandhi's promotion of color television (Pendakur 651), and Egypt's conduct during the Aswan Dam project (Rycroft & Szyliowicz 664). Centralization of R&D management tends to stifle innovation, as is found in the case of dynastic China (Wen-Yuran 950). MacCormack (38) shows how government policy inevitably is enacted without concern for resource efficiency or social justice. African governments promulgate an S&T policy heedless of traditional African technologies, community needs or socio-economic conditions, generally taking cues instead from multinational corporations (Forje 606, Melody 823, Mody & Borrego 901). Government policy promoting the hiring of intellectuals draws a society's most talented individuals from the independent search for scientific knowledge, encouraging them to emphasize instead bureaucratic promotion, as Needham and Ronan (930) find in Chinese history.

Probably the harshest criticism of government intervention in technology policy has been directed at India (Bell & Scott-Kemmis 131). Varghese (696) castigates Indian government for engaging in short-sighted, dependency-inducing technology transfers rather than relying upon educational institutions, research facilities and grassroots science movements to do the S&T work for it. India's import substitution policies are blamed for holding back the potential importation of innovations, and the government's formal R&D organizations are accused of ignoring and indeed drowning out indigenous technological innovations (Deolalikar & Evenson 591). Indian regulation of civil nuclear power is found to be conducted by unaccountable subgovernments (Sharma 678). Indian subsidies for diesel engines hurt the diffusion of biogas engines (Bhattia 357). Nevertheless, the assessment of India's performance has not been all bad. Desai (484) finds that the government is moving toward looser policies that have allowed corporate R&D to surge disproportionately when compared to increases in public projects. By the mid 1980s, the Indian government had retreated from the "commanding heights of the [national] economy" (Bhagavan 568).

Three areas of government intervention pose special interest: (1) telecommunications, (2) military S&T, and (3) trade policy. Governments have special incentive to dabble in the first because of the potential power represented by communications technologies--media access can result in decentralized power (Ogan 905). Pool (831), in a major political science text, argues that new technologies break down national boundaries, threatening the state; government policy in response suppresses diversity and centralizes control of media. Buchner (886), for example, finds that desire for social control results in promotion of television but suppression of telephone technologies; the former lends itself to government domination of communication, but the latter promotes decentralization of media access. Lent (632) claims that desire for increased state power resulted in centralized media and computer industries in Malaysia. Fadul and Straubhaar (888) make similar claims about media facilities in Brazil; centralization ultimately endangers the nation's technological independence. Finally, Sussman (913) applies similar logic to the case in Singapore.

The military is crucial both because of its desire to maintain power in many LDCs and its need to fund research and development in weapons technologies. Dickson (593), for example, finds that military regimes on average spend less on education than other governments. On the other hand, Hess and Mullan (745) see no trade-off between military and educational spending within LDCs, leading one to conclude that the low attention to education by military regimes is a product of low government accountability rather than high arms costs. Gvishiani (966) argues that military concerns tend to dominate Third World scientific research.

Finally, government activism over trade policy is both very common and very likely to distort technology importation as a function of market needs. Clark and Parthasarathi (586) find that border policies on importation can influence S&T development even more than policies establishing a scientific infrastructure. Both Alam (558) and Scott Kemmis and Bell (675), for example, fault Indian protectionism for damping technological development. Cooper (284) also calls for trade liberalization in India as a means to increase high-tech demand. Lee and Lim (630) point to governmental controls aimed at promoting environmental protection as a source of anti-development policy influence. Finally, Pack (331) accentuates the importance of a liberal trade regime for research potential.

International Influences

Treatment of the role played by international forces in LDC development usually seeks signs of exploitation. In particular, the United States is often blamed for interference in the Third World (as Figueiredo 602 argues in the case of Brazil). U.S. laws are isolated as one of the main obstacles to diffusion of science and technology (Weil 701). Although most technology transfer operates under the influence of a foreign-policy context, U.S. transfers especially are influenced by security concerns (Karake 179-180). Sometimes the negative involvement is seen in apparently innocuous programs. For example the Fulbright program may have had the detrimental effect of standardizing technical education (Ilchman & Ilchman 749).

Organizations established to assist the Third World often receive low marks. Foreign aid agencies, for example, are found to encourage Third World governments to utilize foreign experts rather than their own, and to push those governments under the influence of transnational corporations (Mtewa 643, Sussman 912). The few independent attempts at cooperation within the periphery also seldom succeed. Paez-Urdaneta (906), for example, argues that an international information order has failed to develop and urges nations to develop their own information strategies.

Probably the most frequent target for criticism, however, is the global business concern. Third World development basically takes place within a context imposed by transnational (TNC) or multinational (MNC) corporations (Lall 188). Schumann (835) has provided one of the most disparaging discussions. TNCs are seen to aggravate international power imbalances and Third World unemployment rates, as well as to buttress traditional gender roles within LDC cultures. Kenney (820) argues that TNC exploitation is likely to be particularly harsh with the increasing economic importance of biotechnology. However, corporations do not receive universal criticism for their influence on development, since they often fund the training of indigenous managers who later can direct indigenous firms (Gershenberg 1987).

Furthermore, numerous scholars have outlined positive effects stemming from international influences on Third World nations. Amarasuriya (878) argues that UNESCO and similar organizations assist Third World scientists in getting access to necessary information. Moll (902) also suggests that UNESCO has aided in building regional information networks in the periphery. Herzog (922) finds that the best S&T links between the Third World and industrialized nations are formed when native scientists work in foreign or international organizations with peers from the developed nations. Cooperation among Southern Hemisphere countries can improve the likelihood that technologies developed for LDCs will be appropriate to their technical needs (Chiang 580, VanDam 108). Blumenthal (276) argues that international trade has become so important that the nation-state has lost its sovereignty over economic issues; unilateral action on trade policy is unlikely to be effective, so international cooperation is necessary. Finally, even agencies of the U.S. government may not always hurt LDCs (Compton 365).

Scholars seem to be developing a consensus that the international context -- the web of international relations and organizational memberships -- shapes scientific and technological development as much as many national variables (Chisman 720, Ernst 599). Of course, numerous scholars attack the influence of particular organizations or countries on the Third World, as is shown above. However, few scholars, even among those with the more negative evaluations of international influence, suggest that LDCs can ignore most of their international contacts or terminate much of their foreign trade if they hope to increase S&T capacity. Some academics even look to international contacts as the primary means to build up capacity, for example in managing natural resources (Hope 172), funding research and development (Johnston & Sasson 175), and handling nuclear capacity (Luddemann 860).

Institutions

Although the political institutions in a developing nation obviously have the potential to influence the country's path of scientific and technological development, they clearly are not the only ones able to do so (Ahmed & Ruttan 352, Conroy 725). Numerous other institutions hold sway over the shape of S&T acquisition. The most obvious of these is the LDC's educational facilities, both public and, where applicable, private (de Almea 727, Eisemon 734, Ernst 599). But other institutions also have a bearing--for example the professions. Dozens of authors have turned their attentions to the importance of institutional configurations in S&T development.

The most common criticism of educational institutions in the Third World is that they are highly influenced in structure and curriculum by Western values and Western modes of education (Abu-Laban & Abu-Laban 710, Baark 880, Bennell 715, Kwong 761, Mazrui 768, Najafizadeh & Mennerick 777, Selvaratnam 787). Western modes of research organization are seen as poorly matched to Third World needs (Arbab & Stifel 712, Hill 612), although Eisemon (732) argues that African universities only adopt Western models selectively. Or, alternatively, the educational institutions are seen to serve the interests of global capitalism (Bowles 719, Fuenzalida 920), thereby promoting exploitation. Benavot et al. (714) make the more general point that Third World educational institutions tend to vary little in organization, itself problematic regardless of the actual source of the structure. In a discussion the periphery's development, Clignet (723) challenges the need for growth in post-secondary education along any model.

Nevertheless, many scholars see educational institutions as having great potential either to damage or to spur a nation's S&T development (Altbach et.al. 468, Clarke 722, Colclough 724). Kahane (755) suggests that Third World universities do a decent job mediating between LDCs and the developed countries. Furthermore, universities are very important institutions in the recruitment and socialization of the political and scientific elite in many nations (Lomnitz et.al. 517). Universities are important in maintaining personnel needs in crucial S&T fields (Rao 782). Morgan and Armer (733) find that Western-model educational institutions in Nigeria tend to complement traditional educational institutions rather than drive them out. In terms of potential negative effects, Murphy (775) criticizes Third World universities for serving as a mechanism to channel development funds away from utilitarian paths. Weis (796) argues that university funding processes tend to exacerbate inequalities in LDCs through institutional recruitment patterns.

The professions also can have great bearing upon a nation's S&T potential (Eisemon 733, Radwan 780). Third World professions are not seen as organized for fulfilling national needs (Godfrey 1977), assuming professionalization can even take hold as institutions develop (Botelho 576). Meanwhile, educational institutions devote heavy resources to promoting the top layer of professions at the expense of the nation's overall level of development. Third World countries will pour lots of money and effort into educating an elite of engineers and scientists but neglect the sub-professional level of technicians and assistants needed in the country before useful S&T work can be performed (Rao 782, Salam 672; but see Irele 751 for dissent). Left without work requiring application of their skills, the top professionals suffer unemployment difficulties at home (Blaug 718, Meske & de Alaiza 770). Professionals who do get jobs often find their labor wasted by inefficient application (Ukaegbu 793), and have difficulty getting attention from the international scientific community when they publish quality research (Spagnolo 542). The incentive is for these top professionals to emigrate to developed nations, a "brain drain" that results in LDCs subsidizing education for professionals to work in the core countries (Bhagwati 717, Dowty 730, Toh 790), a problem that accelerates when developed countries suffer labor shortages in technical fields (Macphee & Hassan 765).

Scholars have indicated dozens of other institutions that might hurt or help the development process (Forje 608), including bibliographic facilities (Aina 875), libraries and book vendors (Ali 876, Haider 741), publishing firms (Baark 879), mechanisms for technical choice (Basalla 916, Bhatt 572), institutions disseminating S&T information (Brittain 885, Galal 889), a general communications infrastructure (Ho & Sung 893, Moravcsik 772), risk-assessment institutions (Covello & Frey 9), organizations well placed to help workers displaced by technology (Edquist 806), and most importantly, R&D laboratories forming an S&T infrastructure (Arnold 561, Blume 475, Eliou 736, Krishna 509, Langer 897, Lastres & Cassiolato 627, Moravcsik 772, Sagasti 975).

A nation's institutions tend to be evaluated along a number of criteria. One is flexibility or adaptability--the potential for institutions to change with new technologies or development processes (Beranek & Ranis 956, Fischer 603). An institution that is mired in traditional behavior cannot adapt to the stresses placed upon it by new S&T (James & Street 619), yet S&T development often will be more successful if it can operate within existing institutions that already have a place in the society (Glyde & Sa-yakanit 495). Another important structural variable is the quality of technology choice and innovation processes (Bhaneja 570, Bhatt 571, Hoffman 852, Pack 331). A third is centralization -- when the S&T process in a developing country becomes too top-heavy (or "institutionalized") it is likely to be less innovative and less responsive to local needs (Shenhav & Kamens 943). Institutions may be analyzed according to how well integrated they are with other S&T facilities in the country (Crane 10, Sathyamurthy 784), as well as in the international S&T community, since a world scientific community is developing (Schott 535) that offers indigenous scientists and engineers a chance for foreign training (Ukaegbu 792).

The final factor in evaluating S&T institutional performance is more complex than the others: the degree to which the technology choice mechanisms allow feedback from the actual eventual users. That is, do processes within the institution select technology "appropriate" to a society's needs? This factor has mired academics in debate, because of an implicit ideological bent in its initial formulations. The first scholars to champion "appropriate technology" such as E.F. Schumacher and Frances Stewart seemed to give the concept an anti-technological prejudice. That is, to the extent they pinned down exactly what an AT was, they gave the impression that "appropriate technology" was any labor-intensive, "indigenous," inefficient technique (Bruch 63, Eckaus 69). The AT movement was criticized for romanticizing technology choice problems (Ahmad 54).

For all the attacks on the AT movement, however, no true alternative approach counterbalances it. Few scholars are claiming that "bigger is better." Rather, the opposition seems dominated by a free-market ideology--that the "market" will determine what an appropriate technology actually is according to what people are willing to buy and entrepreneurs are able to make succeed. Bigger often will be better because it usually is more productive, but if labor-intensive technologies can be more efficient economically then they may prevail (Bruch 63). However, this faith in the market--no matter how well founded--ignores the impotence of the "market" to decide much of anything in most Third World economies. If the wealth of literature cited in this essay reveals anything, it is that numerous non-market, non-economic factors influence technology choice and technology success in the Third World. These decisions are highly mediated by governmental and corporate institutions that are as insulated from market concerns when transferring technology as any AT laboratory.

Given that market mechanisms will not be making technical choices in the near future, the responsibility seems to lie with government and TNC policies. If so, the arbitration normally provided by consumer decisions in the market must be replaced by a similar institutional process. This possibility for input usually is described as an institution's openness to "feedback." To what extent can institutions making technology choices sense whether the decisions they make are optimal for the users? As Beniger (883) suggests, many institutions assume a one-way flow of information on technology, and are little open to feedback. For example, Bennell (716) finds that universities train engineers according to educational requirements that are not at all shaped by the needs of employers in the Third World. Chatel (64) finds that UN agencies often emphasize purely bottom-line economic concerns in making technology transfer choices, rather than considering likely preferences of target groups. Abiodun (953) argues that African technical choices are not open to possible feedback about African needs; institutions making transfer decisions generally choose to "overgorge" on technology. In contrast to these critiques, Gamser (375-376) provides three examples of institutions that opened their decision processes to feedback from users and were able to engage in "interactive R&D." Rather than stumbling down blind alleys developing products that looked good on paper but were not appropriate to user needs, institutions such as the World Health Organization and Peru's International Potato Center accepted guidance from their "consumers."

Summary

Although Third World poverty helps explain the lag in scientific and technological development found in the Southern Hemisphere, it is not sufficient to explain the variation among nations of the periphery. Scholars increasingly have turned to non-economic factors to explain why some nations in the Third World do better that others. One such factor is cultural. Less-developed countries are seen to possess a pre-industrial mentality, elite ideologies hostile to technology, linguistic barriers to science, and inadequate human resources for S&T development. Also, Western science and technology may contain inherent cultural choices and biases hostile to implantation in Third World cultures. Another non-economic factor is political. Many S&T choices are embedded in governmental institutions heavily vulnerable to political influences and responsive to political needs that have no bearing on optimal S&T choice, including international influences that may not have the LDC's best interests in mind. A final factor is institutional. The shape of a nation's institutions--both scientific and otherwise--may inhibit or promote optimal S&T choice, depending on their adaptability, procedural efficiency, interinstitutional and international linkage potential, and openness to feedback from potential users of technologies to be chosen. While none of these explanations can serve as a replacement for awareness of economic barriers to S&T development in the periphery, they provide a necessary supplement to what otherwise would be a mechanistic understanding of the processes underlying development performance.

D. Stephen Voss

Bibliography

The following works are cited in the above essay, but do not appear in the annotated bibliographic material.

Carroll, Terrance G. 1984. "Secularization and States of Modernity." *World Politics* 36: 362-82.

Gershenberg, Irving. 1987. "The Training and Spread of Managerial Know-How, A Comparative Analysis of Multinationals and Other Firms in Kenya." *World Development* 15: 931-39.

Godfrey, Martin. 1977. "The Third World Professional and Collective Self-Reliance--Towards the Barefoot Accountant?" *IDS Bulletin* 8(3): 19-23.

Henriquez, Pedro. 1983. "Beyond Dependency Theory." *International Social Science Journal* 35(2): 391-400.

So, Alvin. 1990. *Social Change and Development: Modernization, Dependency, and World-Systems Theories*. Newbury Park, CA: Sage.

Chapter 8 -- Politics and Policy

554. Adler, Emanuel. 1986. "Ideological 'Guerrillas' and the Quest for Technological Autonomy: Brazil's Domestic Computer Industry." ***International Organization*** **40: 673-705.**

Asks why Brazil, which has suffered from the classic dependency syndromes, has been successful in implementing a computer policy explicitly aimed at reducing dependency. Finds that domestic institutions gave antidependency "guerrillas" a base for challenging IBM and other MNCs.

555. Adler, Emanuel. 1987. ***The Power of Ideology: The Quest for Technological Autonomy in Argentina and Brazil.*** **Berkeley and Los Angeles: University of California Press.**

Based on case studies of Argentina and Brazil. Both countries faced a choice between a strategy of foreign technological acquisition and one of selected technological autonomy. Argues that a nation's decision can be attributed to the ideologies of actors involved in the decision-making process, as well as their perception of whether the country can set its own economic and technological objectives. This is followed by a study of the role played by ideologically motivated intellectuals.

556. Adler, Emanuel. 1988. "State Institutions, Ideology and Autonomous Technological Development: Computers and Nuclear Energy in Argentina and Brazil." ***Latin American Research Review*** **23: 59-90.**

How can autonomous technological development be explained where conditions offered small potential for it? Comparison of the successful Argentinean nuclear policy and the failed Brazilian policy; comparison of the successful Brazilian computer industry with the failed Argentinean industry. Success is explained by the role of ideological groups and state institutions rather than by structural factors. Ideology of nationalism and antidependency has produced a set of "pragmatic, antidependency guerrillas" (scientists and technocrats) in Latin American countries. The article discusses the use of authoritative technical knowledge to affect decision making in state institutions, creating political shelters, political trust, and political time (a period long enough to create a critical mass of support). Claims that the cases presented help lay to rest explanations of state behavior as the "rational act" of a "rational actor"--what appeared rational to the Argentine Comision Nacional de Energia Atomica (CNEA) appeared irrational to the Argentine hydroelectric lobby, and what seemed rational to the Brazilian government agency CAPRE appeared irrational to political actors who believed in the efficiency of the market.

557. Akindele, R.A. 1976. "Nigeria." ***International Journal of Social Science*** **28: 11-25.**

The first of a four-part series called "Science and Politics: Four National Case Studies." (Others are the U.S., Belgium, and Australia.)

558. Alam, Ghayur. 1988. "India's Technology Policy: Its Influence on Technology Imports and Technology Development." Pp. 136-156 in ***Technology Absorption in Indian Industry*****, edited by Ashok V. Desai. New Delhi: Wiley Eastern Limited.**

Introduction gives a thumbnail history of Indian technology import policy (relatively liberal in 50s and early 60s, increasingly protective from mid-60s). Attempts to evaluate the restrictive technology policy, based on the study of foreign collaborations approved between 1977 and 1983. Uses questionnaires and interviews with 211 technology-importing firms, and discussions with government officials. Finds that there is a large demand for technology in Indian industry and that policies have had only limited success in promoting Indian technological development.

559. Anderson, R.S. 1983. "Cultivating Science as Cultural Policy: A Contrast of Agricultural and Nuclear Science in India." ***Pacific Affairs*** **56: 38-50.**

Argues four main claims about the sciences in India. (1) Following independence the state had assumed an obligatory responsibility for and pre-emptory interest in cultivation of the sciences. (2) The cultivation of different sciences in India spoke to different cultural and social interests which required different treatments. (3) Officials and politicians had to insist on the advantages of science even while some of their own constituents sought to resist its cultivation. (4) Advocates had to account for and adapt to this doubt and resistance because science was at the center of state's other major undertakings.

560. Anthony, Constance G. 1988. *Mechanization and Maize: Agriculture and the Politics of Technology Transfer in East Africa.* New York: Columbia University Press.

Informative, brief volume on the politics of technology transfer in agricultural production, with particular attention to the role of international agencies and state building in Africa. Sensitive to the interaction of politics and technology. The introduction of both tractors and small-scale technology has been marked by limited success and lack of consensus.

561. Arnold, W. 1988. "Science and Technology Development in Taiwan and South Korea." *Asian Survey* 28: 437-50.

Argues that developmental bureaucrats in Taiwan and South Korea have harnessed building of S&T institutions to two principal purposes: (1) the upgrading of industrial structure to catch the West, (2) an increase in military capabilities in face of security dilemma. S&T development is a state activity. While the discovery "push" model is the basis of most science and technology policies, it rarely yields results. South Korea and Taiwan are exceptions. The advancement of S&T are an integral part of economic planning. In Korea, the government played more of an activist and centralizing role than in Taiwan, which established a National Science Council in 1967. Taiwan is more dependent on foreign investment from America and Japan.

562. Baark, Erik. 1981. "China's Technological Economics." *Asian Survey* 21: 977-99.

Examines recent Chinese interest in "realistic" S&T policy through an analysis of the decision to use steam or diesel locomotives. Finds that the lack of feasibility studies prior to development decisions often results in major economic disasters. Suggests that policy makers cannot really pin down an economically rational choice of technology.

563. Bailey, Conner et al. 1986. "Fisheries Development in the Third World: The Role of International Agencies." *World Development* 14: 1269-75.

Shows how Western values impregnated fishing technologies exported to the Third World, and hurt local fishers by promoting capital-intensive, export-oriented processes.

564. Baum, Richard. 1982. "Science and Culture in Contemporary China: the Roots of Retarded Modernization." ***Asian Survey*** **22: 1166-86.**

Even though the Cultural Revolution is over, S&T are constrained by normative and cultural barriers which linger from Confucianism and are "stylistically consonant" with communist ideology: cognitive formalism, narrow empiricism, dogmatic scientism, feudal bureaucratism, and compulsive ritualism.

565. Baum, Richard. 1983. "Chinese Science After Mao." ***Wilson Quarterly*** **7(2): 156-67.**

Discusses a number of strong Chinese traditions that still serve as barriers to scientific progress despite the post-Mao commitment to "modernization." Much of the discussion looks to Chinese culture and philosophical attributes: stereotyped formalism, politicization of science, bureaucratism, and ritualism.

566. Beckler, David. Z. 1992. "A Decision-Maker's Guide to Science Advising." ***Technology In Society*** **14: 15-28.**

Argues that normally helpful S&T advisors can distort decision-making processes and undermine the credibility of their profession unless they are utilized properly. Offers formal guidelines to help in policy formulation. The 13 suggested principles generally emphasize both the need to strengthen the influence of S&T advice and the need to adapt to the subjective, contextual nature of such information.

567. Berlinguet, Louis. 1981. "Science and Technology for Development." ***Science*** **213: 1073-6.**

Science and technology are an integral part of international development. Governments in developed countries must be convinced that money invested in Third World S&T is for the mutual benefit of all.

568. Bhagavan, M.R. 1988. "India's Industrial and Technological

Policies into the Late 1980s." *Journal of Contemporary Asia* 18: 220-33.

Examines India's historical experience with technology, policies, and plans into the late 1980s, including a discussion of the evolving structure of industry in the last two decades. Argues that 1985 signaled the retreat of the state from the "commanding heights of the economy" through (1) simplification of the procedure of licensing industries; (2) increased efficiency of public and private sectors by increasing internal and external competition; (3) modernizing technology through technology transfer from the West.

569. Bhagavan, M.R. 1990. *The Technological Transformation of the Third World: Strategies and Prospects*. London: Zed Books Ltd.

Discusses the global impact of modern technology. Includes discussion of current technological "situations" in the Third World, emphasizing manufacturing, mining and mineral processing, petroleum, agriculture, and the service sector.

570. Bhaneja, Balwant. 1976. "India's Science and Technology Plan 1974-79." *Social Studies of Science* 6: 99-104.

Summarizes the Science and Technology Plan, 1974-79, put together by more than 1,800 Indian scientists and technologists over a two-year period. Explains how this NSTP differs from other, more liberal, plans. Also points to a number of organizational, planning and inceptional constraints under which the plan was formulated.

571. Bhatt, V.V. 1979. "Indigenous Technology and Investment Licensing: The Case of the Swaraj Tractor." *Journal of Development Studies* 15: 320-30.

Shows how systems of technology use sometimes do not give the desired results for LDCs, using a case study of the Swaraj tractor in India. The investment licensing system (ILS) in India was designed to regulate the volume and pattern of investment in the manufacturing sector according to development objectives and strategy worked out by the Planning Commission. However, the ILS decision process was based on poor information and lacked incentives for improving the information base. As a result, the ILS was unable to make adequate demand projections. The author suggests that the "learning process" needs to be institutionalized in the decision-making structure. This may be done through creating a plurality of banks and technical consultancy centers which would compete and thereby goad one another to innovation and act as checks on each other.

572. Bhatt, V.V. 1980. "Financial Institutions and Technology Policy." ***World Development*** **8: 813-822.**

LDCs have a growing amount of scientific and technological knowledge, but poor mechanisms for making technological choices. Offers suggestions for technology policies in Third World countries. Maintains that the problem of poverty must be faced first, by achieving full employment. The crucial problem is to link modern sector development with growth of the traditional sectors, rather than trying to supplant traditional skills and techniques. When making technology choices, this requires upgrading traditional technologies and adopting modern technologies in a manner consistent with development objectives.

573. Bhatt, V.V. 1982. "Development Problem, Strategy, and Technology Choice: Sarvodaya and Socialist Approaches in India." ***Economic Development and Cultural Change*** **31: 85-100.**

Compares two different Indian approaches to technology acquisition--the Sarvodaya perspective of Gandhi and the socialist one of Nehru. Outlines some of the problems in technology choice: upgrading traditional technology, upgrading and improving modern technology in a manner consistent with development objectives, changing institutional structure, and working with natural resources.

574. Blanpied, William A. 1984. "Balancing Central Planning with Institutional Autonomy: Notes on a Visit to the People's Republic of China." ***Science, Technology and Human Values*** **9: 67-72.**

The problems of centralization versus decentralization loom paramount in rebuilding the Chinese S&T base after the Cultural Revolution. Some Chinese leaders believe U.S. science is actually more centralized than science in China because funds for R&D in China are also available at provincial and local levels. Central planning need not imply central implementation in China.

575. Blecher, Marc J. and Gordon White. 1979. "Micropolitics in Contemporary China: A Technical Unit During and After the

Cultural Revolution." ***International Journal of Politics*** **9: 1-135.**

This book-length monograph discusses the impact of the Cultural Revolution on scientists and technicians in China, focusing on the political behavior of a 248-person technical unit in western China. Topics covered include the personnel, structure, and operations of the unit, general issues in the unit from 1966 to 1974, and a narrative account of the Cultural Revolution and its effects. The main problem is that the entire account is based on recollections by one man, albeit one with an encyclopedic memory. Doesn't cover the specific technical activities of the unit, but rather the coalitions and organizational changes over this time period.

576. Botelho, Antonio Jose Junqueira. 1990. "The Professionalization of Brazilian Scientists, the Brazilian Society for the Progress of Science (SBPC), and the State, 1958-60." ***Social Studies of Science*** **20: 473-502.**

The SBPC between 1950 and 1960 tried to define the logic of science according to the professional interests of scientists, but was defeated by the growth of state institutions. Unlike in Western societies, institutionalization was "top-down" while professionalization was "bottom-up" in Brazil. The state's creation of scientific institutions outside the academic context resulted from these opposing logics of political action.

577. Bregman, Arie. 1987. "Government Intervention in Industry: The Case of Israel." ***Journal of Development Economics*** **25: 353-67.**

Examines Israeli government intervention in industry, especially in the capital market, and its effect on rates of return, exports, and the structure of industry. Government intervenes both in direct ways, such as lowering the price of capital, and in indirect ways, such as reducing the risk of investment, expanding human capital, and encouraging R&D. Concludes that government had a substantial, negative effect on industrial performance--industries in which intervention was greater had a smaller rate of return. In particular, the government's policy of encouraging investment is found to cause overinvestment and thereby reduce return. Draws these conclusions from regression analysis.

578. Broaded, C.M. 1983. "Higher Education Policy Changes and

Stratification in China." *China Quarterly* 93: 125-37.

Content analysis of higher education articles in the *People's Daily* in 1971, 1975, and 1978. The general finding is that political/ideological factors were de-emphasized in 1978 after the fall of the Gang of Four. The goal of promoting S&T was strongest in 1978, lowest in 1975. References to S&T curricula highest in 1971. Methodology: seven "destratification" aspects of Chinese higher education were coded and analyzed in terms of number of references per 1,000 lines of text.

579. Chai, T.R. 1981. "Chinese Academy of Sciences in the Cultural Revolution: A Test of the Red and Expert Concept." *Journal of Politics* 43: 1215-29.

China is torn by two competing needs that have always stood in conflict--the need for technical "experts" to promote development and the need for politically reliable generalists ("reds") to foster socialism. This study examines whether redness or expertness was the focus of the Cultural Revolution. Although more "experts" in the Academy faced political attack than "reds", Chai attributes this pattern to political factors shared by the professional class such as party affiliation, age, and party seniority. Once these are taken into account, the real victims of the Cultural Revolution tended to be Reds and "non-real" experts.

580. Chiang, Jong-tsong. 1989. "Technology and Alliance Strategies for Follower Countries." *Technological Forecasting and Social Change* 35: 339-50.

Searches the U.S.-Japan-Taiwan historical experience to find feasible strategies for "follower" countries attempting to adapt outside technologies. Conclusions include: (1) relying on "traditional strategies" reduces adaptability; (2) cooperative national technological systems can accelerate "technological metabolism"; (3) the importance of "process innovation" does not undermine importance of "production innovation"; (4) joint ventures among follower countries are beneficial.

581. Chichilnisky, Graciela and Sam Cole. 1979. "A Model of Technology, Domestic Distribution, and North-South Relations." *Technological Forecasting and Social Change* 13: 297-320.

Outlines a model used for ongoing study of technology policies and applies it to Brazil.

582. Choi, Hyung Sup. 1986. "Science and Technology Policies for Industrialization of Developing Countries." *Technological Forecasting and Social Change* 29: 225-40.

Discusses policies and strategies for S&T development that might be used during the industrialization of a developing country. Industrial growth is the impetus for national development. Draws on the Korean experience.

583. Choi, Hyung Sup. 1987. "Mobilization of Financial Resources for Technology Development." *Technological Forecasting and Social Change* 31: 347-58.

How S&T is financed in the Third World, using Korean example. Direct government financing constitutes the major portion in most developing countries, ranging from 75% to 100%. Most activities are undertaken by institutions in the state sector and financed mainly by the national budget. This may cause problems since S&T projects are generally long-term and budgets are prepared annually and may change. Special funds for science are therefore needed. The state also can play an indirect role through venture capital, incentives, and foreign investments. To use wealth generated by subsidiaries of TNCs, national S&T policies need to be linked more closely with foreign investment policies.

Institutions such as national investment banks need improvements in efficiency and staffing. The productive sector, both public and private, should be brought into the mainstream of research, not simply asked to perform given tasks. With regard to external funding, the UN should prepare a directory of multilateral sources for funding S&T projects. Korea's Science and Technology Development Plan is reviewed. This involved a three-pronged approach to technology development, emphasizing manpower development, accelerated introduction of foreign technologies, and stimulation of domestic R&D.

584. Choi, Hyung Sup. 1988. "Direction for Technological Self-Reliance--Korean Approaches." *Technological Forecasting and Social Change* 33: 23-32.

Focuses on the role of government and industry in creating a favorable environment for strong science and technology in developing countries. Government must link industry and R&D organizations, as well as invest in long-term, high-risk research if dependence on foreign techniques is to be escaped. Uses Korean examples to illustrate strategies. Shows ways beyond mere appropriation through which government can promote development.

585. Choi, Hyung Sup. 1988. "Science Policy Mechanisms and Technology Development in the Developing Countries." ***Technological Forecasting and Social Change*** **33: 279-92.**

Focuses on the role of government and industry in creating a favorable environment for strong science and technology in developing countries. Government must link industry and R&D organizations, invest in long-term, high-risk research, and promote education and cooperation if dependence on foreign technology is to be escaped. Offers ways beyond budgetary appropriation through which governments can promote development, using Korean examples.

586. Clark, N. and A. Parthasarathi. 1982. "Science-Based Industrialization in a Developing Country: the Case of the Indian Scientific Instruments Industry 1947-1968." ***Modern Asian Studies*** **16: 657-82.**

In 1943 a National Planning Committee set up a subcommittee to help establish a viable and self-sufficient scientific instruments industry. Development of the industry in India did not evolve in a manner anticipated by the Committee. In practice, it took a decade for indigenous manufacture to progress, during which period all national requirements were met through imports. Factors influencing rapid growth of production in India were import restrictions and a protected home market, economic growth (especially industrial growth), and gradual development of ancillary industries. The influence of the scientific infrastructure was minimal.

587. Cooper, Charles (ed.). 1973. ***Science, Technology, and Development: The Political Economy of Technical Advance in Underdeveloped Countries.*** **London: Frank Cass.**

Originally published as a special issue of the *Journal of Development Studies*, this early volume contains a nice collection of papers on science, technology, and production in LDCs. Includes pieces on science policy in Latin America, engineering consultancy, skill requirements, and technology choice.

588. Cuthbert. Marlene L. and Stewart M. Hoover. 1991. "Video Beachheads: Communication, Culture, and Policy in the Eastern Caribbean." Pp. 263-278 in *Transnational Communications: Wiring the Third World*, edited by Gerald Sussman and John A. Lent. Newbury Park, CA: Sage.

Maintains that the video revolution is threatening the movement for cultural identity in the Eastern Caribbean. The problem of cultural identity is considered in the historical context of the region. The Caribbean Basin Initiative is considered as a turn away from the regionalism that had emerged in the 1970s. Lack of local government policy initiative is seen as part of the problem.

589. Da Rosa, J. Eliseo. 1983. "Economics, Politics and Hydroelectric Power: The Parana River Basin." *Latin American Research Review* 18: 77-108.

Historical, political, and economic discussion of treaties for and construction of binational hydroelectric dams at Itaipu by Brazil and Paraguay and at Yacyreta and Corpus by Argentina and Paraguay. Looks at prospective problems such as readjustment of royalties and compensation. Useful for those interested in the political background of technological projects. References mostly in Spanish and Portuguese.

590. Dean, G. 1972. "Science, Technology, and Development: China as a 'Case Study.'" *The China Quarterly* 51: 520-34.

Early piece, written during the Cultural Revolution, discussing China's policy of "self-reliance"--accepting technology transfer only when it occurs in response to domestic demands. Particular emphasis on how technologies are chosen, sources of technological change, the role and direction of change, and scientific policy and research priorities. Reports discussions about China's development within the Sussex Study Group on Science and Technology.

591. Deolalikar, Anil B. and Robert E. Evenson. 1990. "Private Inventive Activity in Indian Manufacturing: Its Extent and Determinants." Pp. 233-253 in *Science and Technology: Lessons for Development Policy*, edited by Robert E. Evenson and Gustav Ranis. Boulder: Westview Press.

Attempts to describe the extent of inventive activity in India's manufacturing sector and survey econometric literature on determinants of private inventive activity in India. Determinants of a firm's inventive activity might include firm size, imports of foreign technology, international supply of technology and knowledge, and firm ownership (private ownership vs. state ownership). Concludes that India's import substitution policies have had little success in promoting technological development. Also, firm size is found not to affect research intensity, since most indigenous inventive activity in a country like India is informal and carried out on the shop floor.

592. Desa, V.G. 1978. "Research Co-ordination and Funding Agencies in Developing Countries." ***Impact of Science on Society*** **28: 105-16.**

Advances the National Science Foundation as the desirable model for developing countries to ensure coherent planning and implementation of the national R&D system.

593. Dickson, Thomas, Jr. 1977. "An Economic Output and Impact Analysis of Civilian and Military Regimes in Latin South America." ***Development and Change*** **8: 325-45.**

Finds that military regimes spend less on education than civilian regimes.

594. Dore, Ronald. 1984. "Technological Self-Reliance: Sturdy Ideal or Self-Serving Rhetoric." Pp. 65-80 in ***Technological Capability in the Third World*****, edited by Martin Fransman and Kenneth King. London: Macmillan.**

Marvelous, chatty, sometimes tongue-in-cheek essay (glossary included) focusing on India and the rhetoric of self-reliance in development. Discusses the distinction between learning technology and creating technology as it is embedded in developing country attitudes and personnel.

595. Edquist, Charles and Staffan Jacobsson. 1988. "State Policies, Firm Strategies and Firm Performance: Production of Hydraulic Excavators and Machining Centres in India and Republic of Korea." Pp. 157-208 in ***Technology Absorption in Indian Industry*****, edited by Ashok V. Desai. New Delhi: Wiley Eastern Limited.**

Comparative examination of the effects of state intervention in production. Finds that policies must be chosen according to the nature of particular products and countries. The decision also should be shaped by production objectives; the initiation of production requires state intervention while production for export requires liberalization.

596. Edwards, M. 1989. "The Irrelevance of Developmental Studies." ***Third World Quarterly*** **11: 116-47.**

Argues that the problem with current development studies is the absence of strong links between understanding and action. The first section of the paper considers conventional approaches to development and outlines factors underlying their failure to come to grips with the problems. These factors involve the treatment of people as objects rather than subjects of their own development, the devaluation of popular knowledge, differences between the priorities of researchers and the researched, monopoly of power and knowledge by the North, and the separation of understanding and social change. The second section of the paper offers a model for the future, based on participatory or action research. The author's concept of participatory research is heavily influenced by Paolo Freire.

597. Elliott, David W.P. 1982. "Training Revolutionary Successors in Vietnam and China, 1958-76: The Role of Education, Science and Technology in Development." ***Studies in Comparative Communism*** **15: 34-70.**

Contrasts the Vietnamese and Chinese approaches to technocrats. Vietnam, as a former colony, did not exhibit the extreme egalitarianism and suspicion of technical expertise found in China, although it shares with China the need for organizational settings conducive to harnessing such expertise. Vietnam has been quicker to import technology, a lesson China seems to be learning.

598. Erber, Fabio Stefano. 1985. "The Development of the 'Electronics Complex' and Government Policies in Brazil." ***World Development*** **13: 293-309.**

Traces development of two electronics industries in Brazil (including statistics on their attempts to expand into an international market). Important role of MNCs and local government policy.

599. Ernst, D. 1981. "Technology Policy for Self-Reliance: Some Major Issues." *International Social Science Journal* 33: 466-80.

Suggests that to apply science and technology effectively to key development objectives, developing countries would have to restructure considerably their present international economic, political, and military relations, and thoroughly overhaul their educational systems. Observes that we still lack systematic research on identifying carriers of self-reliance and their conflicting interests.

600. Evenson, Robert E. 1990. "Intellectual Property Rights, R&D, Inventions, Technology Purchase, and Piracy in Economic Development: An International Comparative Study." Pp. 325-355 in *Science and Technology: Lessons for Development Policy*, edited by Robert E. Evenson and Gustav Ranis. Boulder: Westview Press.

Discusses the relationship of intellectual property rights (IPRs) with different stages of development. Includes data on patents and other forms of intellectual property protection in a wide variety of countries. Concludes that IPRs do play a role in development, and that stronger IPRs can help poor countries move forward in technology.

601. Evenson, Robert E. and Gustav Ranis (eds.). 1990. *Science and Technology: Lessons for Development Policy*. Boulder: Westview Press.

A collection of papers concerned with the impact of technology choice and technological change on various aspects of development. The book is divided into five parts. The first part consists of a general introduction with four general observations: (1) import substitution strategies are foolish for developing countries, which should import technologies at the lowest possible cost, (2) policies designed to regulate technology transfer and importation can raise costs and reduce imports, (3) the dominant feature of successful developing countries is a large volume of technology imports, and (4) more economic analysis is needed to guide policy.

The second part contains general theoretical approaches and typologies regarding science and technology in development. The third part offers papers on the Asian NICs and Japan as examples of country-level experience in the use of science and technology for development. The fourth part looks at sector-level experience (agriculture in South and Southeast Asia, industrial efficiency and technology choice in Kenya and the Philippines, and Indian manufacturing). Ends with essays on the process of international transmission of technology in terms of international exchange and of appropriateness and modification of technology.

602. Figueiredo, Nice. 1987. "Informatics in Brazil." *Information Development* 3: 203-7.

Mostly political article that deals with the creation of the National Policy for Informatics in Brazil, and opposition to this policy from the United States. Benefits of the policy are seen as limited by the greater expense of Brazilian products, excessive centralization of the Special Secretariat of Informatics (SEI), lack of control over semiconductor technology, and the unmet need to finance the development of strategic sectors, such as microelectronics.

603. Fischer, William A. 1984. "Scientific and Technological Planning in the People's Republic of China." *Technological Forecasting and Social Change* 25: 189-208.

Excellent discussion by a professor of business administration who spent 2 years in China. Describes the mechanics of the Chinese S&T planning process, including the planning hierarchy and the governmental system (provincial and municipal S&T commissions) that parallels the industrial hierarchy. The governmental commissions can be used for funds if the ordinary vertical (industrial) channels don't work.

Since the planning system involves first the solicitation of expert opinions, then the submission of proposals after the development of a plan, it allows the reaction of enterprises to the plans to drive the system. Since the Cultural Revolution, the lack of people to fill these evaluative positions has sometimes caused all projects to be selected and funded. The main criterion for funding is the financial size of the project. Projects which can be funded by the performing organization won't be reported until relatively complete. A network of contacts can subvert attempts to decentralize the planning system by allowing lower managers access to top levels.

604. Fong, Pang Eng and S. Gopinathan. 1989. "Public Policy, Research Environment, and Higher Education in Singapore." Pp. 137-176 in *Scientific Development and Higher Education*, edited by

Philip G. Altbach, et al. New York: Praeger.

Analyzes industrial research programs and policies being developed by Singapore. Describes key institutions that make and implement science and technology policy. Examines the impact of the research environment in the National University of Singapore on productivity of academics in five scientific fields (biochemistry, botany, computer science, electrical engineering, and physics). Statistical tables on academic research and education in Singapore.

605. Forje, John W. 1979. "Science and Technology: The African Search for a Third Way to Development." Pp. 355-369 in *Science, Technology and the Social Order*, edited by Ward Morehouse. New Brunswick, NJ: Transaction Books.

Asks why both the socialist and capitalist paths to development have failed in Third World countries. Suggests that the answer may lie in the legacy of colonialism. Divides adaptive strategies for African countries into short-term and long-term. Short-term strategies are grouped under three main headings: behavioral change, determination of priorities, and technical operations. Long-term reconstructive strategies must be based on consensus among African nations and long-term development strategies.

Suggests a number of proposals to be discussed at the 1979 Vienna Conference (UNCSTD), including creation of an African science corps, establishment of an African common marker, and joint research ventures. Calls for greater recipient-country control over tech transfer. Discusses the New International Economic Order and the role of S&T in Africa in the changing international system.

606. Forje, John W. 1986. "Two Decades of Science and Technology in Africa." *Science and Public Policy* 13: 89-96.

MNCs and African governments ignore traditional African technologies and fail to take into account different socio-economic backgrounds when exporting modern machinery and technology to Africa. Suggests the need for African scientific/technological self-reliance.

607. Forje, John W. 1988. "In Search of a Strategy for National Science and Technology Policy in Africa." Pp. 229-257 in *Science, Technology, and Development*, edited by Atul Wad. Boulder:

Westview.

Suggests that an articulate and unified S&T policy is needed for the development of the continent. The article looks at policy choices faced by African countries, the climate for development, and social problems affecting development. Finds that a basic change in the balance of political and social forces is needed to achieve science policies that will benefit the majority of people. S&T policy is only part of the required state policy approach.

A unified policy for S&T must involve (1) an assessment of the present state of technological capabilities, (2) strategy formulation in terms of policies, programs, and institutions, together with necessary financial and personnel resources, (3) reassessment of coherence of ends and means, (4) a consensus on the desired mix of appropriate technology and pattern of national capabilities.

Divides African countries into those which have functioning S&T policy-making bodies, those which have such bodies but need to strengthen them, and those which do not have policy-making bodies. Considers problems in the development of human resources. Describes the challenges of the future and Africa's interdependence with the North. Concludes that people should initiate their own development by utilizing available domestic resources.

608. Forje, John W. 1989. "Critique of Technological Policy in Africa." ***Philosophy and Social Action*** **15: 102-12.**

Maintains that Africa's overall development is retarded by low development of human potential, particularly in the area of science and technology. Five indicators of "the chain of technological weakness and low innovative creativity" are given: (1) low level of education at the third level, especially in engineering and technology, (2) poor allocation of financial resources to education and technological activities (table of R&D expenditure in five continents in U.S.-dollar estimates), (3) inadequate or nonexistent institutional infrastructures for SIS and R&D activities, (4) little collaboration between research institutions and the productive sector, (5) little "awareness creation" or dissemination of information on scientific issues. General suggestions for advancement of science and technology, such as the development of national science policies and plans, improvement of machinery and research, and a closer connection of R&D and industry. The author is critical of technology transfer from the developed world and favors increased

South-South cooperation.

609. Guangzhao, Zhou. 1991. "An Example of Science Policy-Making in the People's Republic of China." *Technology in Society* 13: 423-25.

Under China's program of economic reform, the general S&T policy is that "economic development must rely on science and technology, while science and technology should be oriented toward economic development." Refers to differences of opinion in the Chinese Academy of Sciences about its tasks and the contributions it should make.

Reviews basic points of a reform plan formulated by the academy, accepted by scientists, and approved by the government. The reform plan involves: (1) maintenance of a small crack force to conduct basic research and establish key laboratories, (2) establishment of comprehensive research centers to accumulate data on natural resources, the environment, ecology, oceanography, and atmospheric sciences and to enhance scientific analysis for national policy making, and (3) promotion of applied and developmental technological research work and formulation of associations with industrial enterprises for scientific research, development, production, sales, and service to provide technology and products directly to the market.

610. Haas, Ernst B. 1990. *When Knowledge Is Power: Three Models of Change in International Organizations*. Berkeley: University of California Press.

Although the subject of the book is internal change among international organizations, the conceptual framework relies heavily on "consensual knowledge" produced by epistemic communities in producing this change. Three models of change are developed, two of which involve adaptation (incremental growth and turbulent nongrowth) and one expressing "learning," a rare condition resulting in basic problem redefinition. A set of descriptive and analytic dimensions is applied to the post-WWII international and regional organizations. The three models of change are compared and suggestions for future designers of organizations are offered.

611. Hill, Hal. 1983. "Choice of Technique in the Indonesian Weaving Industry." *Economic Development and Cultural Change* 31:

337-54.

Looks at different weaving techniques available to Indonesia and finds that wise economic decisions generally were made in choosing. If capital-intensive technologies were chosen, it was the capital and labor policies of government, rather than industry, that was responsible for the choice.

612. Hill, Stephen. 1987. "Basic Design Principles for National Research in Developing Countries." *Technology in Society* 9: 63-73.

Questions the applicability of Western models of research organization to the Third World. The main properties of the research environment in developing countries are "knowledge poverty" and external forces driving change. National research can never play more than a marginal role in development, but it can fill gaps where international interests do not concern themselves. (1) Research must be targeted strategically to utilize very scarce resources and (2) knowledge bridges must be constructed from research to the economy.

613. Hiraoka, Leslie S. 1985. "Japan's Technology Trade." *Technological Forecasting and Social Change* 28: 231-41.

Looks at Japan's historical development from a developing to developed nation, stressing educational and technology transfer policies.

614. Hobday, Michael. 1985. "The Impact of Microelectronics on Developing Countries: The Case of Brazilian Communications." *Development and Change* 16: 313-40.

Outlines the broad industrial and product trends in the world telecommunications market. Examines Brazil's experiences with telecommunications and assesses the government's performance in establishing: (a) local industrial and technological capacity in digital technology, and (b) an efficient, modern telecommunications infrastructure. Concludes by pointing to some wider implications of relevance to other DCs.

615. Inkster, Ian. 1989. "Appropriate Technology, Alternative Technology, and the Chinese Model: Terminology and Analysis." *Annals of Science* 46: 263-77.

A study of China's technology policy since 1949. Argues that institutional changes geared to the absorption of outside advanced technologies are a sign of the success of past technological strategies rather than a denial of them.

616. Jacobsson, Staffan. 1984. "Industrial Policy for the Machine-Tool Industries of South Korea and Taiwan." *IDS Bulletin* 15(2): 44-9.

Discusses how industrial policy in Korea and Taiwan is helping machine-tool firms improve their technological capabilities and move into the production of numerically controlled machine tools. Gives a brief outline of the character of global change in the industry, and shows how the nature of competition has been changing. Gives an outline of the main characteristics of the machine-tool industries in Korea and Taiwan, and shows how the governments have historically encouraged the industry.

Finds that Korea's industry has surpassed Taiwan's in dollar amount of production because of the success of the government's industrial policy. The two main policy causes of Korea's comparative success were trade restrictions and credit. The Korean government has allowed import restrictions to be applied for machine tools that can be produced domestically, and has channeled larger amounts of capital to the machine-tool industry than the Taiwanese government. Thus, Korea had greater success than Taiwan due to greater government intervention.

617. Jacobsson, Staffan. 1985. "Technical Change and Industrial Policy: The Case of Computer Numerically Controlled Lathes in Argentina, Korea and Taiwan." *World Development* 13: 353-70.

Analyzes an effort by these three countries to produce electronic tools, one of the few advanced products that NICs have produced successfully. Gives particular emphasis to the role of government policy in enhancing competitiveness of lathe producers in the NICs. Analyzes the international market for lathes and concludes that the type of lathes the NICs have specialized in producing is the one most affected in the substitution process. Maps barriers to entry for production of CNC lathes. Argues that firm-specific, custom-designed government policies are the most efficient.

618. James, Dilmus D. 1980. "Mexico's Recent Science and Technology Planning: An Outsider Economist's Critique." *Journal*

of Interamerican Studies **22: 163-93.**

Why Mexico attracts students of economic development: (1) it is one of the prime cases of distinction between economic growth and economic development, (2) it offers a test of the dependency claim that the subjugation of the periphery to external control varies according to the length and intensity of contact with the center, (3) economic interest in Mexico as an energy supplier, (4) the problem of illegal migration and the uniqueness of its border economy and (5) Mexico's accelerating S&T program, pushed primarily through the rapid growth of Concejo Nacional de Ciencia y Trabajo (CONACYT).

Mexico faces two salient socioeconomic difficulties--unemployment and distribution of wealth--but the role of S&T in dealing with these problems is not given adequate attention in the National Plan for Science and Technology of November 1976. Offers an overview of this plan and a critique of it. Mexican firms are competitive with foreign firms in obtaining technology and are beginning to show some success at exporting it. The Mexican government can exercise influence over the direction of S&T. The National Plan was superseded by Lopez-Portillo's National Program for Science and Technology.

619. James, Dilmus D. and J.H. Street. 1983. "Technology, Institutions and Public Policy in the Age of Energy Substitution: The Case of Latin America." ***Journal of Economic Issues*** **17: 521-8.**

Events in Latin America are a belated effort to catch up technologically with a development process based on assumptions about energy that are themselves undergoing rapid change. In most instances, traditional institutions are inadequate to cope with new technological requirements. Clearly defined programs for applying scientific effort to the solution of energy problems simply do not exist.

620. James, Jeffrey. 1980. "Appropriate Technologies and Inappropriate Policy Instruments." ***Development and Change*** **11: 65-76.**

Argues that existing approaches to appropriate technology are either incorrect or incomplete, because they lack clear criteria for determining whether a technology is appropriate given any particular context. In particular, two weaknesses in the development literature are blamed for this failure. The first is found in A.K. Sen's work on the choice of techniques. Sen overlooks the problem of political feasibility in implementation of technology, and concentrates on growth rather than the reduction of unemployment or inequality. The second flaw shows up in Schumacher's conception of "intermediate technology." Schumacher places excessive emphasis on "dual economy" countries, thereby ignoring the inequalities perpetuated by traditionally rural sectors. Modern economic sectors receive disproportionate blame for inequalities in his work.

621. James, Jeffrey. 1989. *The Technological Behaviour of Public Enterprises in Developing Countries*. New York: Routledge.

A volume of essays examining specific public enterprises in various countries: petroleum in Argentina, fertilizer in Brazil, and ammonia in Mexico. Also includes general discussions of the market behavior of state-owned enterprises in Tanzania and India, and articles to put public enterprises in theoretical perspective.

622. Jamison, Andrew and Erik Baark. 1991. *Technological Innovation and Environmental Concern: Contending Policy Models in China and Vietnam*. Lund, Sweden: Research Policy Institute.

Based on research in China and Vietnam in 1989-90. This collection of papers examines the tension between a market-oriented technology policy and one intended to protect the natural environment, described as "economic policy culture" versus the "academic policy culture." In China and Vietnam, representatives of both cultures are seeking to transform bureaucratic socialism into a more democratic policy process. Includes a historical overview, the development of an "environmental consciousness," and remarks on innovation policy.

623. Jimenez, J., M.A. Campos, and J.C. Escalante. 1991. "Distribution of Scientific Tasks Between Center and Periphery in Mexico." *Social Science Information* 30: 471-820.

Examines the distribution of scientific tasks between Mexico City (the center) and the Mexican States (the periphery). This includes an examination of the concentration of science and technology in Mexico, and a description of the history and goals of Mexican S&T policy from its beginnings under President Cardenas (1935) to the present. This policy was based both on technological self-determination and on decentralization. Suggests that the more recent institutions, built away from the metropolitan area, are more attuned to official S&T policies because their establishment was predicated on adherence to these policies.

624. Just, Richard E. and David Zilberman. 1988. "The Effects of Agricultural Development Policies on Income Distribution and Technological Change in Agriculture." ***Journal of Development Economics*** **28: 193-216.**

Economic model predicting the effects of policies on income distribution when farmers have discrete adoption decisions but constantly make decisions regarding land allocation. Equity effects depend on farm size, risk preferences, and credit availability. Fixed-cost subsidy is recommended from an equity standpoint in most cases.

625. Kuang-tou, Chang. 1977. "Polytechnic Education in the New China." ***Impact of Science on Society*** **27: 437-42.**

Outlines the principles and goals of training young engineers in China. Of interest because it was written just after the purge of the Gang of Four. Still refers to the fitness of engineer "both for intellectual and manual work." Applied science students are recruited from workers, peasants, and soldiers with 2-3 years experience. An "open door" policy of interchange between students/teachers and workers/peasants. Politics is studied along with math and physics for the purpose of raising socialist consciousness.

626. Kueh, Y.Y. 1985. "Technology and Agricultural Development in China: Regional Spread and Inequality." ***Development and Change*** **16: 547-70.**

Less-developed provinces have managed to catch up with more advanced provinces in the last three decades, reducing regional inequality in agricultural technology. The Chinese government's central planning has sought to squeeze agricultural surplus from more advanced provinces to finance industrialization. Poor provinces have been mobilized to support their own productivity, but numerous subregional inequalities persist. Agricultural policy is to exploit the rich without supporting the poor.

627. Lastres, Helena M. and Jose E. Cassiolato. 1990. "High

Technologies and Developing Countries: The Case of Advanced Materials." ***Materials and Society*** **14: 1-10.**

The development of advanced materials is part of a greater economic and technological revolution that facilitates the creation of new products and new economic sectors. Since the new materials are technology-intensive they tend to make the traditional comparative advantages of developing countries meaningless. This forces countries to strengthen their technological development, to build up their institutional S&T networks, and to develop long-term S&T policies.

628. Lavakare, P.J. and J.G. Waardenburg (eds.). 1988. ***Science Policies: An International Perspective--The Experience of India and the Netherlands.*** **London: Pinter.**

Contains papers from the Indo-Dutch Workshop on Science Policy held in New Delhi. Topics include Indian science policy and national goals, Indian national policies toward R&D systems, and practical uses of the results of the R&D system in India.

629. Lecraw, D.J. 1981. "Technological Activities of Less-Developed Country Based Multinationals." ***Annals of the American Academy of Political and Social Sciences*** **458: 151-62.**

Looks at technology exports by India, the leading exporter of industrial technology in the Third World. Traces this status mainly to the strategy of the Indian government to encourage technological learning in the capital goods sector of the economy. While this protectionist strategy has created inefficiencies, it has also led to the creation of substantial technological capability.

630. Lee, Joon Koo and Gill Chin Lim. 1983. "Environmental Policies in Developing Countries." ***Journal of Development Economics*** **13: 159-73.**

Environmental policies restricting the importation of polluting industries hurt Third World nations economically.

631. Lee, Renssalaer W., III. 1982. "Political Absorption of Western Technology: The Soviet and Chinese Cases." ***Studies in Comparative Communism*** **15: 9-33.**

Outlines the complications involved when importing technology into communist nations, using China and the USSR for comparisons.

Importation is more politicized in China, which hinders the ability of established organizational structures to assimilate change.

632. Lent, John A. 1991. "Telematics in Malaysia: Room at the Top for a Selected Few." Pp. 165-199 in *Transnational Communications: Wiring the Third World*, edited by Gerald Sussman and John A. Lent. Newbury Park, CA: Sage.

Examines Malaysia's information infrastructure with regard to telecommunications, computerization, and mass communications. Looks at who owns, controls, and benefits from communications initiatives. Discusses Malaysia's rapid economic advances during the 1980s, Prime Minister Mahathir's New Economic Policy, and privatization in Malaysian telecommunications.

Shows how heavy investment has accelerated telematics in Malaysia and outlines the progress of computerization. Gives a critical view of Mahathir's restructuring of the mass media, which has made the media more supportive of the government. Concludes that, as a result of the restructuring of the 1980s, the new telecommunications and computer industries are controlled by a few groups and individuals.

633. Leonard, D.K. 1987. "The Political Realities of African Management." *World Development* 15: 899-910.

Addresses importation of inappropriate management technologies to Africa, caused by poor understanding of socio-political contexts. Uses the U.S. Agency for International Development's "Africa Bureau Development Management Assistance Paper" of 1984 to illustrate. Considers policy making, leadership, general internal administration, and bureaucratic hygiene. Stresses special constraints on managerial performance in Africa, especially patronage in the political systems.

634. Macioti, Manfredo. 1978. "Technology and Development: The Historical Experience." *Impact of Science on Society* 28: 313-20.

Author was Director of Science Policy for the Commission of the European Communities. The experience of today's industrialized nations should not be dismissed as irrelevant to the developing world. Features most relevant for the Third World today are the government as innovator and selectivity in technology imports. Japan as example.

635. Mackenzie, Donald. 1984. "Marx and the Machine." *Technology and Culture* 25: 473-502.

Argues that Karl Marx's "account of the machine" is still historically relevant, for example in its application to the "alternative technology" movement. Attempts to dismiss the perception that Karl Marx was a strong technological determinist. Argues that, in large sections of *Das Kapital*, social relationships are described as molding technologies during the development of large-scale mechanized production.

636. Martinussen, John. 1988. *Transnational Corporations in a Developing Country: The Indian Experience*. Newbury Park, CA: Sage Publications.

Reviews changes in India's regulatory policies completed by 1974 and their effects on TNCs, with a view toward policy recommendations for the future.

637. McCulloch, Rachel D. 1981. "Technology Transfer to Developing Countries: Implications of International Regulation." *Annals of the American Academy of Political and Social Science* 458: 110-122.

Discusses the drive within UNCTAD to establish an international code of conduct governing North-South technology transfer. Article examines motives for and probable consequences of an international code. Proposed UNCTAD code would separate tech transfer from other trade and investment transactions--preferential treatment of Southern nations to promote rapid growth and ensure equity in economic relations with the North.

638. Moravcsik, M. J. 1980. "What Motivates the Developing Countries To Do Science and Technology?" *Communication and Cognition* 13: 237-47.

Suggests that developing countries are driven more by a desire for equality with developed countries, in some abstract competitive sense, than by purely material motivations.

639. Moravcsik, Michael. 1981. "Mobilizing Science and Technology for Increasing the Indigenous Capability in Developing Countries." *Bulletin of Science, Technology and Society* 1: 355-77.

A review of the major theoretical and empirical concerns in the academic and popular debate over scientific development. Includes a number of specific recommendations for research and development, communication, education, organization and management.

640. Morehouse, Ward. 1976. "Professional Estates as Political Actors: The Case of the Indian Scientific Community." *Philosophy and Social Action* 2: 61-95.

An excellent description of the Indian scientific elite, its relative lack of autonomy, and the political forces that shape it. The scientific elite is a group of about 200 people who are occupants of important organizational and administrative positions and who link agencies and committees. Since only in the areas of agriculture and nuclear energy has science and engineering been relatively successful, the elite lacks both resources and autonomy. C.P. Snow's types of "closed politics" used as a conceptual framework.

641. Mtewa, Mekki (ed.). 1990. *International Science and Technology: Philosophy, Theory and Policy*. New York: St. Martin's Press.

Editor founded the Association for the Advancement of Policy, Research and Development in the Third World in 1981. This book is divided into five sections. The first is an overview by the editor of the theory and practice of science and technology for development. Sections Two through Five contain articles on educational, institutional, communication, and economic policy.

642. Mtewa, Mekki. 1990. "Science and Technology for Development: Theory and Practice." Pp. 3-26 in *International Science and Technology: Philosophy, Theory and Policy*, edited by Mekki Mtewa. New York: St. Martin's Press.

A review of policy and theory on S&T in development. Looks at what constitutes development and development policy, examines technical utilization and science utilization in developing countries, and reviews theoretical aspects of science and technology policy. This last section offers six theoretical modes within which STD policy is viewed.

In discussing development policy, the author attempts to define goals, objectives, and needs and to consider how these relate to and differ from each other. In the discussion of technical utilization, the author examines how policies affect the maintenance and use of technical professionals. In his discussion of science utilization, the author offers suggestions for the use of science in developmental problem solving.

The examination of theoretical aspects suggests that STD policies are seen from six basic perspectives: (1) systems theory, which sees political systems in terms of input and output processes, (2) elite theory, which sees STD policy as a reflection of the preferences and values of the governing elite, (3) group theory, which sees policy as the product of interactions among groups, (4) rational theory, which assumes that the costs and benefits of policies can be analyzed, (5) developmental theory, which views STD policy as directly related to levels of sophistication and confidence STD policy makers acquire in the use of tools and policies, and (6) language theory, which evaluates biases, ideologies and standards in STD arguments. The author concludes that a development-oriented government with coordinated programs and efficient processes for generating and evaluating research information, with clear and coherent STD policy goals can achieve a cost-effective STD infrastructure. Without these qualities, STD policy formulation and generation is likely to be erratic.

643. Mtewa, Mekki. 1990. "Experts, Advisers and Consultants in Science, Technology and Development Policy." Pp. 75-117 in *International Science and Technology: Philosophy, Theory and Policy*, edited by Mekki Mtewa. New York: St. Martin's Press.

Maintains that developing country governments must understand the role of experts, advisers and consultants in order to use them effectively. Poses two fundamental problems: (1) nearly every development plan formulated by bilateral and multilateral experts has failed, and (2) in order to affect development activities a structure is needed in which permanent government employees and paid consultants can work together. The article deals with comparability of objectives of consultants and clients, and technical assistance objectives of such organizations as the World Bank, UNDP, the Operational and Executive Personnel to Overseas Territories program (OPEX), and the Commonwealth Fund.

Offers a critique of the assistance offered by such organizations. Suggests that African professionals are underutilized because governments and universities continue to allow control by foreign expertise, institutions, and money. Concludes that developing countries should not abdicate their political mandates to foreign experts. A clear distinction of activities should be made between user clients and experts: local professionals should be trained as counterparts to foreign experts in order to translate the professional exchange into development for the client country.

644. Natarajan, R. 1987. "Science, Technology and Mrs. Gandhi." *Journal of Asian and African Studies* 22: 232-49.

By 1966, when Indira Gandhi took power, India already had: (1) a collection of state-sponsored agencies whose activities were not directly related to the production system, (2) an industrial system heavily dependent on foreign technology, (3) government decision-making machinery which lacked guidelines and criteria for monitoring and evaluating costs and appropriateness of imported technology, and (4) infrastructure investments in technology education which had not yet produced sufficient qualified manpower.

Changes during Gandhi's regime: (1) creation of new agencies for research and administration in S&T, (2) expansion in overall funding for existing agencies with changes in relative allocations among agencies, (3) restrictive and selective policy towards foreign investments and collaborations, (4) expansion and revamping of decision-making and consultative machinery for S&T and technology imports, (5) efforts to link S&T system with other production sectors of the economy through the S&T plan, (6) formulation of a technology policy. Advances were made in spite of political infighting and bureaucratic conflicts. Gives a table of R&D expenditures.

645. Nelson, Richard R. 1990. "On Technological Capabilities and Their Acquisition." Pp. 71-80 in *Science and Technology: Lessons for Development Policy*, edited by Robert E. Evenson and Gustav Ranis. Boulder: Westview Press.

Discusses features of modern technology, and what is involved to command technology for an enterprise in a newly industrializing nation. Considers the role of an industrial R&D organization in advancing technology. Looks at the role of the U.S. and the "catching up" of other countries during the seventies and eighties. Examines industrial development problems of the NICs and issues of public technology policy.

646. O Cinneide, M.S. 1987. "The Role of Development Agencies in Peripheral Areas with Special Reference to Udaras na Gaeltachta." *Regional Studies* 21: 65-8.

Discusses the role of development agencies in remote rural regions through a case study of an agency devoted to promoting industrialization in the region of Ireland where the Irish language is still the vernacular, known as the Gaeltacht. Concludes that this agency, the Udaras na Gaeltachta, has helped to arrest the demographic decline of the Gaeltacht through the provision of industrial employment.

647. Odhiambo, Thomas R. 1992. "Designing a Science-Led Future for Africa: A Suggested Framework." *Technology In Society* 14: 121-30.

An optimistic account of Africa's potential for future development, stressing scientific progress and the need for partnerships among business, government and the scientific community.

648. Pal, S.K. and B. Bowonder. 1978. "High Technology and Development in India." *Futures* 10: 337-41.

Offers possible applications of electronics in India and policy alternatives for the Indian government.

649. Parrott, Bruce. 1983. *Politics and Technology in the Soviet Union*. Cambridge: MIT Press.

Historical work on the development of ideas of technological progress and strategy in the Soviet Union, beginning with the Stalinist foundations, through WWII and the Khrushchev years, and ending with Brezhnev.

650. Patel, Surendra. 1984. "Technology in UNCTAD: 1970 to 1984." *IDS Bulletin* 15(3): 63-6.

Assesses the success of this UN agreement in getting Third World nations to include local technological development on their policy agendas.

651. Pendakur, Manjunath. 1991. "A Political Economy of Television: State, Class, and Corporate Confluence in India." Pp. 234-262 in *Transnational Communications: Wiring the Third World*, edited by Gerald Sussman and John A. Lent. Newbury Park, CA:

Sage.

How political and economic forces in India have shaped television there. History of Indian TV since the arrival of black-and-white TV technology in 1955. Argues that the shift in TV policy from education to mass entertainment resulted from political exigencies. Similarly, the development of color TV in 1982 is seen as an attempt to bolster Indira Gandhi's prestige. Concludes that the nationally owned TV network was put at the joint service of the corporate sector and the state to promote the profits of the former and the power of the latter.

652. Perkins, F.C. 1983. "Technology Choice, Industrialisation and Development Experiences in Tanzania." ***Journal of Development Studies*** **19: 213-43.**

Suggests that most of the output in 10 industries studied came from small-scale technological production practices, but finds that the government purchases capital-intensive, less-efficient technologies, owing to budgeting procedures and other factors.

653. Pray, Carl E. and Vernon W. Ruttan. 1990. "Science and Technology Policy: Lessons from the Agricultural Sector in South and Southeast Asia." Pp. 179-207 in ***Science and Technology: Lessons for Development Policy*****, edited by Robert E. Evenson and Gustav Ranis. Boulder: Westview Press.**

Reviews evidence of the effectiveness of past technology policies in Asia, discusses technology policies that might help Asian governments meet agricultural challenges, and draws technology policy lessons for other sectors of LDC economies. Offers a broad range of empirical evidence on R&D in these countries, looks at rates of return to research in Asia and percent of total area of crops devoted to HYV wheat, ice and maize. Offers suggestions on how the Green Revolution may be sustained. Concludes that public-sector research can be a high payoff investment, that the most effective public research systems are closely linked with users of technology, that international centers for topics other than agriculture can be an effective way of increasing scientific and financial resources to address Asian development problems, and that private-sector research can be encouraged with pretechnology R&D by the government and with universities that produce local scientists and engineers.

654. Puryear, Jeffrey M. 1982. "Higher Education, Development Assistance, and Repressive Regimes." ***Studies in Comparative International Development*** **17 (2): 3-35.**

Discusses the negative role authoritarian regimes play in the suppression of research, with focus on Chile, Argentina and Uruguay.

655. Qian, Wen-yuan. 1985. ***The Great Inertia: Scientific Stagnation in Traditional China.*** **London: Croom Helm.**

Written by a physicist who claims China's intolerant political structure inhibited scientific thought. Others who have discussed China's S&T policy have not considered that the "great organizational achievement" represented by the Chinese state promoted technology but smothered science.

656. Rahman, A. 1979. "Science and Technology for a New Social Order." Pp. 317-333 in ***Science, Technology and the Social Order,*** **edited by Ward Morehouse. New Brunswick, NJ: Transaction Books.**

Points out the current limitations of science, problems with the nature of contemporary technology (with regard to control of resources, preservation of inequalities, inefficient machines and systems, pollution), and the unequal relationship between developed and developing countries. Suggests that major transformations in S&T result when S&T developments are linked to a movement of social change.

657. Ramamurti, Ravi. 1985. "High Technology Exports by State Enterprises in LDCs: The Brazilian Aircraft Industry." ***Developing Economies*** **23: 254-80.**

Examines the particularly successful experience of Brazil in promoting exports of a sophisticated manufactured product, light aircraft, using a state-owned enterprise. Attempts to identify country-specific, industry-specific, and firm-specific factors that led to this success, in order to derive policy conclusions for other LDCs seeking to form state-owned enterprises that will develop competitive high-tech industries. Suggests that state enterprises can be successful exporters of high-tech goods, only if a large number of rare preconditions are met. Preconditions include availability of human capital commensurate with technological ambitions of the state venture, a large home market, an institutional design for the enterprise that provides managerial autonomy and goal clarity, and the type of manager who has the combination of motivations and skills necessary for effectiveness.

658. Ramanathan, K. 1988. "Evaluating the National Science and Technology Base: A Case Study on Sri Lanka." *Science and Public Policy* 15: 304-20.

Outlines a framework for gaining clear understanding of a nation's development infrastructure. Goes beyond allocation patterns of R&D expenditures and distribution of S&T personnel to embrace information on critical issues related to the technological attributes of national production system and the scientific/technological climate.

659. Randolph, Robert H. and Bruce Koppel. 1982. "Technology Assessment in Asia: Status and Prospects." *Technological Forecasting and Social Change* 22: 363-84.

Technology Assessment attracted attention in Japan, peaked in 1978, then waned. TA activities were scarce in other countries. Examination of Indonesia, Japan, Korea, Philippines, Taiwan, India, and Iran through interviews in 1980. Activities related to TA have occurred (implicit TA in national planning). Asia is interested in accelerated adoption, not assessment.

660. Ranis, Gustav. 1990. "Science and Technology Policy: Lessons from Japan and the East Asian NICs." Pp. 157-178 in *Science and Technology: Lessons for Development Policy*, edited by Robert E. Evenson and Gustav Ranis. Boulder: Westview Press.

Examines Japan and the East Asian NICs to determine the causes of appropriate technology choices and an appropriate direction of technology change, with regard to both production processes and product quality. Concludes that S&T policy is still hampered by vested interest groups in much of Asia, that the early import substitution phases of the NICs and Japan were less severe than other countries and allowed them to move toward more market-oriented policies, and that without a move toward more open markets the demand for more appropriate technologies cannot assert itself.

661. Rawski, T.G. 1980. *China's Transition to Industrialism: Producer Goods and Economic Development in the Twentieth*

***Century*. Ann Arbor: University of Michigan Press.**

Studies the contribution made by domestic producer industries to China's development. Surveys the nation's producer industries from 1900-37, a period of significant but limited import substitution. Also discusses expansion and diversification in the Chinese private sector promoted by Japanese-backed industries in Manchuria. Considers the rapid expansion of industry during 1949-57 and thereafter. Examines how China's producer sector could succeed in shifting priorities from quantitative toward qualitative considerations within a Soviet-style institutional framework. Looks at historical and international comparisons and considers the extent to which China's successful pattern of industrial and technical development may be relevant to other developing countries.

662. Rushing, Francis W. and Carole Ganz Brown (eds.). 1990. *Intellectual Property Rights in Science, Technology, and Economic Performance: International Comparisons*. Boulder, CO: Westview.

This book stems from a project co-sponsored by the National Science Foundation's International Division. The articles included cover issues in intellectual property rights, international comparisons among developing countries (three of the five articles are concerned with Brazil, the fourth with India, and the fifth with Southeast Asia), and international comparisons among developed countries. Concludes with a piece by Alden F. Abbott which argues that intellectual property protection plays an essential role in inducing R&D, summarizes the constitutional underpinnings and principal forms of protection in the U.S., and discusses possible government initiatives for enhancing intellectual property protection.

663. Rycroft, Robert W. 1983. "International Cooperation in Science Policy: The U.S. Role in Macroprojects." *Technology in Society* 5: 51-68.

Suggests that American dominance in world science no longer holds in all fields and that the U.S. therefore needs to follow cooperative science policies. Gives some of the possible problems in a division-of-labor approach to scientific cooperation and suggests ways to solve those problems. Delineates three promising policy options: (1) developing a database that will define the international scientific division of labor, (2) facilitating National Science Foundation leadership in cooperative efforts, and (3) employing a line-item in each agency budget for international scientific activities.

664. Rycroft, Robert W. and Joseph S. Szyliowicz. 1980. "The Technological Dimension of Decision Making: The Case of the Aswan High Dam." ***World Politics*** **33: 36-61.**

Case study showing how political factors can prevent selection of optimal technologies in LDCs, guided by theories of decision making. Explores the international debate surrounding construction of the Aswan High Dam in the Nile Valley. Finds that both Egyptian and World Bank officials followed an "organizational-process model" in which satisficing rather than optimizing seemed the dominant strategy. All actors operated with bounded rationality, utilizing extremely limited information. In particular, Egyptian leaders made no effort to maximize their choice of technologies once they found one that satisfactorily met political needs for speed and monumental attraction.

665. Sagasti, Francisco R. 1976. "Technological Self-Reliance and Cooperation among Third World Countries." ***World Development*** **4(10/11): 939-46.**

Calls for international cooperation among Third World nations to enhance technological development. Outlines possible cooperative development strategies, then ends with a proposal for formation of an international association composed of LDCs, governed by a small central staff, empowered to identify and nurture S&T projects for member countries.

666. Sagasti, Francisco R. 1978. ***Science and Technology for Development: Main Comparative Report of the Science and Technology Policy Instruments Project.*** **Ottawa: International Development Research Centre.**

One of the largest comparative studies in developing countries supporting the idea of "intimate connections" between production activities, financial and economic decisions, and technology. Links between research and production in LDCs are weak or nonexistent. Policies on foreign investment, credit and interest rates, patent and trade regulations, imports and exports, project analysis criteria, market protection, and social inequity are more influential in determining the direction of technical change than R&D policies.

667. Sagasti, Francisco. 1979. "Towards an Endogenous Scientific and Technological Development for the Third World." Pp. 301-316 in *Science, Technology and the Social Order*, edited by Ward Morehouse. New Brunswick, NJ: Transaction Books.

Examines five components of a strategy for endogenous scientific and technological development: (1) autonomy of decision making in science and technology, (2) the necessity of identifying problem areas where the process of endogenization of the S&T revolution should be pursued, (3) redistribution of the world scientific and technological effort, (4) granting Third World countries privileged access to the acquisition of technologies to satisfy human needs, and (5) cooperation among Third World countries.

668. Sagasti, Francisco R. 1989. "Science and Technology Policy Research for Development: An Overview and Some Priorities from a Latin American Perspective." *Bulletin of Science, Technology, and Society* 9: 50-65.

A detailed look at the research situation in Latin America, including the evolution of development and policy concepts. After outlining expected economic, social, cultural, political, and S&T trends in Latin America, the author proposes four initiatives that might improve science and technology: (1) take better stock of the situation in policy research; (2) promote and support research efforts; (3) provide assistance to teaching and training programs; (4) disseminate research results to the general public.

669. Saich, Tony. 1986. "Linking Research to the Productive Sector: Reforms of the Civilian Science and Technology System in Post-Mao China." *Development and Change* 17: 3-34.

Sinologist from Leiden examines the evolution of S&T policy since 1976, especially finance and personnel. The Ten-Year Plan and frustrated ambitions. Between 1965 and 1978, spending on civilian research was halved, but resources are now on the increase. Problems getting researchers to move to areas where skills are needed.

670. Saich, Tony. 1989. *China's Science Policy in the 80s.* Manchester, England: Manchester University Press.

Examines China's progress in developing its indigenous S&T sector, in particular its S&T research system, arguing that Western writers who look at technology transfer have ignored China's reforms in the S&T sector such as granting more autonomy.

671. Salam, Abdus. 1988. "What the Third World Really Needs." *Bulletin of the Atomic Scientists* 44(9): 8-10.

This editorial offers proposals to address the world's S/T imbalance, such as: (1) a means-related international levy toward research on global problems and (2) a consortium of developed-country universities to enhance science around the world. Finally the author makes suggestions on how agencies set up for international aid could best distribute funds.

672. Salam, Abdus. 1991. "A Blueprint for Science and Technology in the Developing World." *Technology In Society* 13: 389-404.

Argues that any blueprint for upgrading science and technology in the Third World must have two components: (1) a pattern of development in science and technology in these communities that fits their present transitional phase and (2) strategies and mechanisms. Offers broad historical background explaining the technological gap among nations and points to education as a means of closing it.

The general pattern for development of S&T in developing countries should consist of: (1) promotion of science education, (2) structuring basic scientific and technological research in universities to make it a self-reinforcing system and producing technical cadres to support researchers, (3) strengthening the supportive infrastructure by building up expertise in low and medium technologies, emphasizing craftsmanship and fabrication techniques, (4) drawing up a comprehensive plan for applied sciences, and (5) focusing on training of personnel for R&D in high tech. Table at the end of the article gives comparative expenditure on science and technology versus proposed funding for science and high technology for 25 developed capitalist countries, 12 socialist countries, and 98 developing countries.

673. Salam, Abdus and Azim Kidwai. 1991. "A Blueprint for S&T in the Developing World." *Technology In Society* 13: 389-404.

Stresses the importance of technological capabilities in determining economic development, then points with concern to the widening North-South gap in technology. Nations have become technologically and economically interdependent -- with, for example, economic conditions in Brazil affecting the global rainforest supply. This interdependence underscores the need for an international technology policy.

674. Sardar, Ziauddin. 1977. *Science, Technology and Development in the Muslim World.* London: Croom Helm.

Discusses the Muslim view of development, outlining debates within Muslim science on science policy, cultural and ethical dimensions, a social side of development, foreign and trade, agricultural versus industrial priorities, the importance of technological self-reliance, basic or applied R&D, and academia.

675. Scott-Kemmis, Don and Martin Bell. 1988. "Technological Dynamism and Technological Content of Collaboration: Are Indian Firms Missing Opportunities?" Pp. 71-104 in *Technology Absorption in Indian Industry*, edited by Ashok V. Desai. New Delhi: Wiley Eastern Limited.

Based on interviews with British technology suppliers to India. Examines the technological content of Indo-British collaboration agreements. Concludes that in the 1980s there were gaps (1) between suppliers' knowledge resources and the potential technological content of knowledge flow, (2) between the potential technological content and the planned technological content, and (3) between the planned technological content and the actual technological content. Argues these gaps were due to government import policy as well as to other factors.

676. Sen, Asim. 1979. "Followers' Strategy for Technological Development." *Developing Economies* 17: 506-28.

Traces Japan's historical development from an LDC to an economic power, emphasizing development approaches.

677. Sharif, M.N. and V. Sundararajan. 1983. "A Quantitative Model for the Evaluation of Technological Alternatives." *Technological Forecasting and Social Change* 24: 15-30.

Use of model to evaluate Thailand's selection of a railway system.

678. Sharma, Dhirenda. 1983. *India's Nuclear Estate*. New Delhi: Lancers.

Transfer of civilian nuclear power to India. Argues that it is regulated by a subgovernment unaccountable and unregulated owing to non-democratic traditions.

679. Sherwood, Robert M. 1990. *Intellectual Property and Economic Development*. Boulder, CO: Westview.

Author is an international business counselor on intellectual property processes. Considers intellectual property protection systems as a form of economic infrastructure, strong in developed countries and weak in developing countries. Points to a connection between intellectual property protection and innovation, needed for economic development. Good overview of the five basic forms of intellectual property--trade secrets, patents, copyrights, trademarks, and mask works.

Contains case studies of Brazil and Mexico, based on interviews and questionnaires administered to businesspeople, lawyers, and others concerned with intellectual property rights issues. Refutes four arguments in favor of weak protection--that weak protection saves the country money, promotes local industry, helps acquire technology, and lessens dependency--using an empirical examination of developing countries with and without strong protection. Gives suggestions for developing countries. Bibliography is useful and particularly strong on World Bank materials.

680. Shishido, Toshio. 1983. "Japanese Industrial Development and Policies for Science and Technology." *Science* 219: 259-64.

Two important factors contributing to Japan's economic success were government investment in industrial development and the early recognition that a good educational system is a prerequisite to tech progress. Policies promoted the importation of technology from Europe and North America and encouraged the education of students abroad. This facilitated the rapid development of Japanese industry and the adaptation of technology to local conditions. Japanese advantages in technological development were: (1) a centralized political system and sense of unity, (2) utilization of the home market, (3) accumulated capital, (4) coexistence of modern and traditional industries, (5) high standards of education, (6) selection of appropriate technology.

681. Sigurdson, Jon. 1980. *Technology and Science in the People's Republic of China*. Oxford: Pergamon Press.

Outlines China's recent commitment to technological development, as expressed in policy changes, establishment of S&T commissions, educational and intellectual changes and R&D expenditures. Looks in particular at activities in such areas as the basic and applied sciences, agriculture, medicine, electronics, waste management, pest control, and energy.

682. Sinclair, Craig. 1988. "Science and Technology in Greece, Portugal and Turkey." *Science and Public Policy* 15: 354-56.

Shows how NATO cooperation is expanding beyond the military sphere to include attempts at increasing R&D expenditures.

683. Soberon, G. and V. Urquidi. 1992. "The Case of Mexico." *Technology In Society* 14: 131-6.

Outlines policy changes made in Mexico during the late 1980s to spur lagging S&T growth, including a national program for modernization and establishment of a scientific research advisory council for the Mexican executive branch.

684. State Science and Technology Committee of the People's Republic of China. 1989. *Guide to China's Science and Technology Policy*. Oxford: Pergamon Press.

Detailed book with sections on reform of the management system, policy and legislation, "soft" science research, R&D (with statistical indicators), environment and resources, as well as legal policy and regulation. Reprints a few pertinent government documents.

685. Straubhaar, Joseph. 1989. "Television and Video in the Transition from Military to Civilian Rule in Brazil." *Latin American Research Review* 24: 140-54.

Brazil's media-state relations can only be understood in the light of the country's corporatist heritage, which is mixed with populist manipulation of media. Offers a good description of television in Brazil and presents one of the few attempts to treat developing country media in theoretical terms.

686. Subramanian, S.K. 1987. "Planning Science and Technology

for National Development: The Indian Experience." ***Technological Forecasting and Social Change*** **31: 87-102.**

Drawing lessons from the Indian development experience, this paper points to a number of factors essential for planning science and technology to complement levels of national development: improvement of engineering, work ethic, flexibility, creativity, development of human capital along with technological capital. Planning must emphasize results rather than procedures.

687. Surprenant, Thomas T. 1987. "Problems and Trends in International Information and Communication Policies." ***Information Processing and Management*** **23: 47-64.**

Examines the New World Information and Communication Order (NWICO), World Administrative Radio Conference (WARC), and Transborder Data Flow (TBDF) to consider problems in development of international information flow. Author fears that these show a pattern of mutual antagonism that may lead to an "information war." Fairly extensive references on the subject of international information exchange.

688. Suttmeier, Richard P. 1974. ***Research and Revolution: Science Policy and Social Change in China.*** **Lexington, MA: Lexington Books.**

Important volume documenting policy shifts and their effects on Chinese scientific development from the beginning of the socialist regime, through the First Five-Year Plan, the Great Leap Forward and the Cultural Revolution. Stresses the organization of research and the administration of R&D during periods of both "mobilization" and "regularization," but the book was written before the end of the Cultural Revolution.

689. Suttmeier, R.P. 1980. "Science Policy and Organization." In ***Science in Contemporary China*****, edited by Leo Orleans. Stanford: Stanford University Press.**

Nice overview of science policy in China after the end of the Cultural Revolution, including a readable description of the organization involved.

690. Suttmeier, Richard P. 1980. ***Science, Technology, and China's***

***Drive for Modernization.* Stanford: Hoover Institution.**

An account of developments in Chinese science policy since the fall of the Gang of Four and Zhou Enlai's doctrine of the Four Modernizations. Includes a discussion of the linkage between the politics of science and political succession, institutional changes since 1977 (Chinese Academy of Sciences, higher education, production ministries), the planning system, manpower, expenditures, and international relations. Concluding chapter on the politics of Chinese science and Sino-American relations.

691. Suttmeier, Richard P. 1981. "Politics, Modernization and Science in China." *Problems of Communism* 30: 22-36.

An overview of the record of S&T development in post-Mao China. Examines issues of institutional reform, education and manpower, the National Research Plan of 1978, and international relations. Argues that there are serious problems of policy implementation and that these are linked to widespread resistance to institutional change.

692. Turner, Bryan S. 1987. "State, Science, and Economy in Traditional Societies: Some Problems in Weberian Sociology of Science." *British Journal of Sociology* 38: 1-23.

Employs Weber's economic sociology to identify the close historical relationship between economic change, state regulation, and the patronage of intellectuals in the development of science. Scientific rationalism is the outcome of contingent features, structural arrangements between the economy and the state, the presence of rational technologies, and the teleological impact of rationalization. Examples from China, Islam, and the West.

693. Turner, Terisa. 1976. "The Transfer of Oil Technology and the Nigerian State." *Development and Change* 7: 353-90.

Shows how socio-political structures prevent the importation of oil technology. Questions the assumption of UNCTAD literature that poor country governments are seriously striving to obtain technology on the best possible terms. Suggests that technology transfer is first and foremost a political problem and therefore considers the role of the state in transfer. Part I discusses the interests of oil companies. Part II examines the nature of the Nigerian state. Part III reviews the Nigerian experience in transfer of exploration and production technology, downstream project technology, and oil-related expertise.

694. Turner, Terisa. 1977. "Two Refineries: A Comparative Study of Technology Transfer to the Nigerian Refining Industry." *World Development* 5: 235-56.

Finds that Nigeria has changed its means of acquiring technology, with an increase in local participation and control. The result is more appropriate technology, but higher costs (attributable to state ownership and inflation). The second project also was geared to increasing local science and technology development, because locals were involved in construction as well as operation of the refinery.

695. UNESCO. 1986. *Comparative Study on the National Scientific and Technological Policy-making Bodies in the Countries of West Africa*. Paris: UNESCO.

Reports results of a study of the structure, administration and resources of S&T policy-making bodies in 15 West African states. Reports individual results for each nation's policy-making body, followed by an intercountry assessment for the subregion.

696. Varghese, N.V. 1986. "Education, Technology and Development: An Indian Perspective." *International Social Science Journal* 38: 117-25.

Criticizes past patterns of technology transfer into India as shortsighted and dependency inducing. Offers an "innovation chain" method of technology importation, linking educational institutions, basic and applied research facilities and grassroots science movements, intended to replace indiscriminate technology importation with long-term, coordinated development.

697. Vien, Nguyen Khac. 1979. "The Scientific and Technical Revolution in the Socialist Republic of Vietnam." *Impact of Science on Society* 29: 241-46.

Sets out the reasons for Vietnam's technological backwardness and outlines a "great development project" designed to enable the Socialist Republic of Vietnam to become an economically self-sufficient country by the year 2000. The overall management of scientific and technical research is entrusted to the National Scientific Research

Center.

698. Vogel, Ezra F. 1991. *The Four Little Dragons: The Spread of Industrialization in East Asia.* Cambridge: Harvard University Press.

East Asian scholar seeks to explain the interaction of government and industry in South Korea, Taiwan, Singapore, and Hong Kong. Each used some version of the Japanese model, but each is different: Korea is most similar to Japan; Taiwan is oriented toward entrepreneurs and social network ties; Singapore plans every facet of life. Ethos of "industrial neo-Confucianism" is used to describe bureaucrats. Each government picks strategic directions to promote and has a sense of shared national purpose.

699. Volti, Rudi. 1982. *Technology, Politics and Society in China.* Boulder: Westview Press.

The history of major technological policies in China since 1949, ideological orientations, and organizational patterns. Case studies of ground transportation, agriculture, energy, medicine and public health. Interviews with technicians in technology transfer.

700. Wad, Atul. 1984. "Science, Technology and Industrialisation in the Third World." *Third World Quarterly* 6: 327-50.

Argues that social scientists fail to understand problems with Third World development, and emphasizes the role of politics. Good historical background (most of the statistics are derived from other sources).

701. Weil, Vivian. 1988. "Policy Incentives and Constraints on Scientific and Technical Information." *Science, Technology and Human Values* 13: 17-26.

Includes discussion of U.S. government's use of law to restrict dissemination of scientific and technological data outside U.S.

702. Weiss, Charles, Jr. 1985. "The World Bank's Support for Science and Technology." *Science* 227: 261-5.

Outlines the many projects sponsored by the World Bank to promote the development of AT in LDCs. Discusses how budgetary woes have forced the reevaluation of priorities. Includes a good bibliography of articles studying World Bank-affiliated projects.

703. Westphal, Larry E., Kopr Kritayakirana, Kosal Petchsuwan, Harit Sutabutr, and Yongyuth Yuthavong. 1990. "The Development of Technological Capability in Manufacturing: A Macroscopic Approach to Policy Research." Pp. 81-134 in *Science and Technology: Lessons for Development Policy*, edited by Robert E. Evenson and Gustav Ranis. Boulder: Westview Press.

A research study, conducted under the auspices of the Thailand Development Research Institute, centering on three areas: biotechnology, electronics and information technology, and materials technology. Attempts to answer the following questions: (1) how well do Thai key industries utilize technology? (2) how does government policy affect utilization? (3) what new policies or policy changes could improve utilization? Technology utilization is evaluated through technology capability scores awarded by teams of economists and technologists after visits to plants.

704. Whiston, Thomas G. 1988. "Co-Ordinating Educational Policies and Plans with Those of Science and Technology: Developing and Western Developed Countries." *Bulletin of the International Bureau of Education* 247: April-June.

Describes methods to coordinate educational policy with rapid changes in science and technology. Good annotated bibliography.

705. Wignaraja, Ponna. 1979. "Science and Technology in a New Development Strategy." Pp. 335-353 in *Science, Technology and the Social Order*, edited by Ward Morehouse. New Brunswick, NJ: Transaction Books.

Examines the top-down framework of development, which depends on transfer of external capital and technology, and attempts to formulate an alternative perspective. Argues that techno-economic ideas of development do not contribute to social change or to development in a human sense. The new alternative perspective is said to require new processes and a non-alienating technology.

706. Wionczek, Miguel S. 1981. "On the Viability for a Policy on Science and Technology in Mexico." *Latin American Research Review* 16: 57-78.

Underlying idea is that Mexico does not have a scientific/technological policy. Examines why the National Plan for Science and Technology presented in November 1976 failed. The Comprehensive Development Plan of May 1980 gives only cursory attention to science and technology problems.

707. Wise, Steve. 1990. "Space and National Development: Are Brazil and Argentina Examples?" *Technology In Society* 12: 79-90.

Examines the space programs of Brazil and Argentina, the leaders in Latin American space development. Provides a history of the two countries' national and international space activities. Concludes that in order to profit from space technology programs, developing countries must find ways to develop supporting areas, such as communications or teacher training. Space technology does not allow a nation to bypass stages in technological development, but can help speed up the process if in proper combination with other systems. A developing country's political and social strategy is even more important in determining how a space program will influence development. Brazil and Argentina can offer examples for other countries by dispelling the notion that the benefits of a space program are out of reach for developing countries, and because they have chosen space technology appropriate to their own needs and means.

708. Yuthavong, Yongyuth et al. 1985. "Key Problems in Science and Technology in Thailand." *Science* 227: 1007-11.

Defines high priority problems applicable to Thailand's scientific/technological development: the availability of human resources, cutting-edge advantages to be gained, the possibility of international collaboration, integration with cultural environment, and relevance to development. The authors recommend Thailand focus on bioscience, material science, electronics, and information science.

709. Zahlan, A.B. 1980. *Science and Science Policy in the Arab World*. New York: St. Martin's Press.

An in-depth look at Arabic science and technology, especially thorough for Egypt, with emphasis on scientific human and financial resources, the international and regional dimension of science, and formal science policy. Argues that the Arab nations are more committed to science than is commonly perceived.

Chapter 9 -- Education and the University

710. Abu-Laban, Baha and Sharon McIrvin Abu-Laban. 1976. "Education and Development in the Arab World." ***Journal of Developing Areas*** **10: 285-304.**

Western models have often been imposed upon educational institutions. A "third" level of education discussed.

711. Ajibero, Matthew Idowu. 1985. "Attitudes of Faculty Members Toward Media Technologies in Nigerian Universities." ***British Journal of Educational Technology*** **16: 33-42.**

Based on 250 questionnaires distributed to university faculty members in Nigeria. Finds that the majority of respondents showed a positive attitude toward media technology. The extent to which individual values will affect the actual use of these technologies in instruction remains to be determined.

712. Arbab, Farzam and Laurence D, Stifel. 1982. "University for Rural Development: An Alternative Approach in Colombia." ***Journal of Developing Areas*** **16: 511-22.**

Examines a university that succeeded by deviating from the Western mode of university organization.

713. Ayala-Castanares, Agustin. 1983. "The Role of Universities in Building National Capability in the Marine Sciences." ***Impact of Science on Society*** **33: 405-11.**

Gives the Institute of Marine Science and Limnology (ICML) at the Mexican Autonomous National University (UNAM) as an example of the evolution of scientific research. Initial efforts were devoted to manpower training, both in foreign universities and locally, followed by establishment of a graduate program. Now significant numbers of graduate students are finishing their doctorates abroad. A great deal of the development is attributed to international cooperation.

714. Benavot, Aaron, Yun-Kyung Cha, David Kamens, John Meyer, and Suk-Ying Wong. 1991. "Knowledge for the Masses:

World Models and National Curricula, 1920-1986." *American Sociological Review* 56: 85-100.

Primary school curricula are very similar throughout the world, owing to expansion of the nation-state system and the dominance of standardized models of mass education; similarity is not predicted by existing theories, which suggest cultural differences.

715. Bennell, Paul. 1983. "The Professions in Africa: A Case Study of the Engineering Profession in Kenya." *Development and Change* 14: 61-81.

Informative article describing how Anglo-style professionalism and university structure persisted after Kenya achieved independence, excluding qualified technicians from certain fields of labor.

716. Bennell, Paul. 1986. "Engineering Skills and Development: The Manufacturing Sector in Kenya." *Development and Change* 17: 303-24.

Examines the results of research into determinants of the demand for skilled engineering personnel in the manufacturing sector in Kenya. The discussion is based on two main propositions: (1) the level of effective demand for formally qualified personnel in the manufacturing sector of Kenya is relatively limited; (2) the pattern of demand for engineering personnel bears little resemblance to requirements defined in relation to formal educational/vocational qualifications.

717. Bhagwati, Jagdish. 1976. "The Brain Drain." *International Social Science Journal* 28: 691-729.

Article reviews existing evidence on flow of skilled manpower from developing to developed countries, looks at consequences of brain drain, and considers proposals to deal with this phenomenon. Proposes a tax on immigrants' incomes in developed countries, to be routed to developing countries through the UN.

718. Blaug, Mark. 1979. "Economics of Education in Developing Countries: Current Trends and New Priorities." *Third World Quarterly* 1: 73-83.

Discusses educational planning policies and the unemployment of Third World academicians and professionals.

719. Bowles, Samuel. 1978. "Capitalist Development and Educational Structure." ***World Development*** **6: 783-96.**

Theoretical overview of the relationship between the accumulation process and educational change in the capitalist periphery. Argues that educational systems may serve capitalist class interests in four ways: (1) by regulating the labor flow between capitalist and noncapitalist means of production, (2) by raising productivity in the capitalist mode, (3) by thwarting the development of class-conscious proletariats or peasantries, and (4) by undermining the positions of the traditional (landed) elites. Offers data to support the contention that educational systems in these countries primarily serves capitalism.

720. Chisman, Dennis G. 1984. "Science Education and National Development." ***Science Education*** **68: 563-9.**

Looks at the relationship between science education and national development in a very general and summary fashion. Describes the Vienna Conference on Science and Technology for Development (UNCSTD) in 1979 and the UNESCO Congress on Science and Technology Education and National Development in 1981. Suggests that science and technology education may be seen as taking place within national development or for national development (i.e., the education receiving its form and purpose within the general social framework versus education being seen as an attempt to influence national development through strategic use of science and technology).

721. Choudhuri, Arnab R. 1985. "Practicing Western Science Outside the West: Personal Observations on the Indian Scene." ***Social Studies of Science*** **15: 475-505.**

Discusses problem of adapting modern science for non-Western regions. Extremely important hypothesis on teaching science in graduate school: countries on the periphery emphasize breadth of training and classical ideas about science, but not actual research apprenticeship.

722. Clarke, Ronald. 1977. "Extending the University for Grass-Roots Development." ***IDS Bulletin*** **8: 46-51.**

Examines the potential for Third World universities as agents for development.

723. Clignet, Remi. 1977. "Educational and Occupational Differentiation in a New Country: The Case of the Cameroun." ***Economic Development and Cultural Change*** **25: 731-46.**

Challenges the idea that growth of post-secondary education is important to economic growth.

724. Colclough, Christopher. 1982. "The Impact of Primary Schooling on Economic Development: A Review of the Evidence." ***World Development*** **10: 167-85.**

For the Third World, a large impact of primary schooling.

725. Conroy,R.J. 1989. "The Role of Higher Education Sector in China's Research and Development System." ***China Quarterly*** **117: 38-70.**

Examines the changing role of higher education in the Chinese R&D system. The author sees a concerted effort to mobilize and apply the S&T resources of the higher education sector for the sake of economic development. Rapid expansion of higher education and its growing centrality to R&D in China. Reforms in higher education are part of wider reforms and these appear to have had an impact only after 1984. Still, a relatively small group of established, well-staffed institutions is doing a disproportionate amount of R&D.

726. D'Ambrosio, Ubiratan. 1979. "Knowledge Transfer and the Universities: A Policy Dilemma." ***Impact of Science on Society*** **29: 223-30.**

Looks at the role universities must play in Third World development, emphasizing Latin America. Suggests that technology transfer should be considered a temporary, first stage in the construction of a country's scientific and technical establishment.

727. de Almea, Ruth Lerner. 1977. "Innovation in the Harnessing and Transfer of Technology: The Gran Mariscal de Ayacucho Foundation." ***Impact of Science on Society*** **27: 299-307.**

Describes an experimental model for the training of Latin American students, both at home and abroad. Concludes that this program for the harnessing and transfer of science both reinforces the need for cooperation and reaffirms the dependence of national development on human resources.

728. Dietz, James L. 1990. "Technological Autonomy, Linkages, and Development." Pp. 177-200 in *Progress Toward Development in Latin America: From Prebisch to Technological Autonomy*, edited by James L. Dietz and Dilmus D. James. Boulder: Lynne Rienner Publishers.

Suggests a technological strategy for Latin America based on technological autonomy. Institutionalist theory is used to argue for the importance of education and human-resource skills. Elites may work to frustrate this strategy while paying lip service to it.

729. Dorozynski, Alexandre. 1976. "Science, Technology and Education on the Arabian Peninsula." *Impact of Science on Society* 26: 193-9.

Special issue on science in the Islamic world. This article maintains that the Arabian peninsula is gradually finding its way into a contemporary world in which science and technology play leading cultural roles. Rapid modernization of traditional educational processes is contributing to this evolution.

730. Dowty, Alan. 1986. "Emigration and Expulsion in the Third World." *Third World Quarterly* 8: 151-76.

Discussion of emigration of professionals and scientists from Third World countries. Brain drain statistics.

731. Dugan, Kathy. 1987. "History of Science in Non-Western Classrooms: A Bridge Between Cultures." *Social Studies of Science* 17: 145-61.

Discusses cultural assumptions in Western texts, especially applied to New Guinea and China.

732. Eisemon, Thomas Owen. 1980. "African Academics: A Study of Scientists at the Universities of Ibadan and Nairobi." *Annals of the American Academy of Political and Social Sciences* 448: 126-38.

How Western academic and intellectual traditions are being transformed at two African universities. Both are described as faithful to the traditions of Western universities in many ways, but these traditions are selectively modified. The academic and intellectual culture of these universities contains traditions which are indigenous and unreconciled with their colonial heritage.

733. Eisemon, Thomas Owen. 1980. "Scientists in Africa." *Bulletin of Atomic Scientists* 36: 17-22.

Nigeria has the largest and most developed scientific community in Africa. Kenya was slower and less aggressive in developing science. Looks at why. Also points to a growing disillusionment and skepticism among scientists in both nations regarding their own profession, although differences exist between the generation of scientists who entered scientific careers during the transition to independence and those that began afterwards. Kenya's universities still rely on expatriates more than Nigeria's. Nigerian scientists, meanwhile, are frustrated by the gap between the realities of practicing science and the way scientific administrators in the country perceive that process.

734. Eisemon, Thomas Owen. 1982. *The Science Profession in the Third World*. New York: Praeger.

A comparative study of science in India and Kenya, looking at its historical development in both countries. Gives special emphasis to the structure of higher education in both nations--autonomy, authority, the evolution of scientific careers, bureaucratization, hierarchy.

735. Eisemon, Thomas O. and Yakov M. Rabkin. 1978. "Science in a Bilingual Society: The Case of Two Engineering Schools in Quebec." *Social Studies of Science* 8: 245-56.

Although not on the Third World, the ideas here are relevant to bilingual Third World nations.

736. Eliou, Maria. 1981. "Research in Higher Education in Greece." *Journal of the Hellenic Diaspora* 8: 123-30.

Discusses limited resources for Greek R&D and training of younger faculty outside Greece.

737. Forje, John W. 1990. "The Management of Science and Technology for Industrial Development in Africa: The Case of Cameroon." *Journal of the Third World Science, Technology, and Development Forum* 8: 234-44.

Need for more cooperation between institutes of higher education and manufacturing industries to achieve self-reliant, sustainable industrial development. Appropriate endogenous technologies need to be incorporated in industrialization.

738. Gareau, Frederick H. 1986. "The Third World Revolt against First World Social Sciences: An Explication Suggested by the Revolutionary Pedagogy of Paulo Freire." ***International Journal of Comparative Sociology*** **27: 172-89.**

Looks at the process whereby Third World academics reject First World paradigms and adopt their own.

739. George, June M. 1988. "The Role of Native Technology in Science Education in Developing Countries: A Caribbean Perspective." ***School Science Review*** **69: 815-20.**

A short discussion of the benefits gained by students in developing countries when native technologies are incorporated into the science curriculum. Stresses intangible benefits such as increased feeling of self-worth, illustrating her point with the example of the steelpan musical instrument as a way of teaching sound-wave physics in Trinidad/Tobago.

740. Grabe, William and Robert B. Kaplan. 1986. "Science, Technology, Language and Information: Implications for Language and Language-in-Education Planning." ***International Journal of the Sociology of Language*** **59: 47-71.**

Authors seek to determine the relation between scientific information and language (in particular English) and to examine what the science/language relation implies for aspects of Human Resources Development Planning. These issues serve to define the roles of English and Chinese in modernization efforts.

Modernization requires freedom from dependence on technologically advanced countries. To reduce dependency on outside technology requires an information-access system which improves R&D--this requires access through English. The information system must have (1) a large number of English users, (2) an accessible, up-to-date library system, (3) scholarly journals, (4) a sophisticated computer retrieval network, (5) a large network of informal contacts.

741. Haider, Syed Jalaluddin. 1989. "Acquisition of Scientific Literature in Developing Countries. 3: Pakistan." ***Information***

***Development* 5: 85-98.**

The third in a series of articles on the acquisition of scientific literature by libraries in developing countries. Summarizes the background, the development of science and technology in Pakistan, library and information system development, and the existing library resources in science and technology.

742. Handberg, Roger. 1986. "Practising Western Science Inside the West: Psychological and Institutional Parallels between Western and Nonwestern Academic Cultures." *Social Studies of Science* 16: 529-33.

Extends Choudhuri's article about the problems of conducting Western science in a non-Western culture by pointing out that some of the obstacles are also problems in the West for people trained at elite universities but working in lesser institutions.

743. Hayhoe, Ruth E.S. 1987. "China's Higher Curricular Reform in Historical Perspective." *China Quarterly* 110: 196-230.

Changes in the structure of curricular knowledge after 1911 were faced with persistence of traditional Chinese Confucian traditions (lack of disciplinary boundaries, pure knowledge). The European model of university organization, with centralized control, was adopted by Nationalists in mid 1930s, but the actual system was closer to American. Contradictions grew in each period up to 1966. The Cultural Revolution closed universities, then created links between universities and production units. Higher education reforms of 1985 show reemergence of traditional patterns, but the emergence of social science is considered problematic.

744. Herschede, Fred. 1980. "Chinese Education and Economic Development: An Analysis of Mao Zedong's (Mao Tse-tung's) Contributions." *Journal of Developing Areas* 14: 447-68.

Historical analysis of the success of Mao's educational policies, connecting education and economic development.

745. Hess, Peter and Brendan Mullan. 1988. "The Military Burden and Public Education in Contemporary Developing Nations: Is There a Tradeoff?" *Journal of Developing Areas* 22: 497-514.

Focuses on the determinants of military expenditures in less-developed countries, examining the relationship between spending on defense and public education. Much of the article is devoted to the discussion of regression methodology. The simple two-equation model for the shares of public expenditures on education and the military on GNP found no trade-off or substitution effect.

746. Hewson, Mariana G. and Hamlyn Daryl. 1985. "Cultural Metaphors: Some Implications for Science Education." *Anthropology and Education Quarterly* 16: 31-46.

Explores the question of whether cultural metaphors concerning heat are still dominant among Sotho people in Southern Africa and what implications this metaphor has for learning orthodox scientific theories. The metaphors, entailing "prekinetic" notions of heat, persist. Suggests that educators use these metaphors to communicate the contemporary scientific kinetic view of heat.

747. Holland, Susan S. 1976. "Exchange of People Among International Companies: Problems and Benefits." *Annals of the American Academy of Political and Social Science* 424: 52-66.

Exchanges of people by international companies leads to improved understanding, but this depends on the participants' ability to understand and adapt. This ability can be developed by adequate training.

748. Holsinger, Donald B. and Gary L. Theisen. 1977. "Education, Individual Modernity, and National Development: A Critical Appraisal." *Journal of Developing Areas* 11: 315-34.

Finds that education fuels technological development by creating personal malleability, needed for technology transfer.

749. Ilchman, Warren F. and Alice Stone Ilchman. 1987. "Academic Exchange and the Founding of New Universities." *The Annals of the American Academy of Political and Social Science* 491: 48-62.

Includes discussion of role of Fulbright program in Third World and similarities among international curricula. Other articles in this issue look at the program's specific effects on Latin America, Africa, etc.

750. Ingle, Henry T. 1986. "New Media, Old Media: New Technologies of International Development." *International Review of Education* 32: 251-67.

Research, theory, and practice over past 75 years provides convincing evidence that educational technology offers a comprehensive and integrated approach to solving educational and social problems. Effective use of these technologies requires a long-term commitment to financial support and training of personnel.

751. Irele, Abiola. 1990. "Education and Access to Modern Knowledge." In *A World to Make: Development in Perspective*, edited by Francis X. Sutton. New Brunswick, NJ: Transaction.

Maintains that the modern ascendancy of the West is largely due to the development and cultivation of the scientific spirit, and that Western science is to be seen as a mode of cultural expression. Based on this position, the article examines education and its bearing on development in Nigeria, assumed to be representative of Africa in general. Nigerian educational planning is said to be more appearance than substance, resulting in the mass production of semiliterates. Emphasis on quantity of students has been destructive of quality and has stretched resources, resulting in a shortage of facilities. In order to contribute to development, education should reject the model of quantitative expansion in favor of the cultivation of a frame of mind attuned to the scientific model.

752. Jairath, Vinod K. 1984. "In Search of Roots--The Indian Scientific Community." *Contributions to Indian Sociology* 18: 109-30.

Looks at why a strong scientific community has failed to develop in India despite a large quantitative growth in scientific personnel, investment, and institutions. Suggests that the problem is ultimately political, and that it can only be solved by a political solution based on vast social changes, to solve the isolation and fragmentation of Big Science.

753. Jamison, Dean T. and Peter R. Moock. 1984. "Farmer Education and Farm Efficiency in Nepal: The Role of Schooling, Extension Services, and Cognitive Skills." *World Development* 12: 67-86.

Uses data from Nepal to ascertain the relation between education and farmer efficiency. Finds that education has a positive effect for three major crops (early paddy, late paddy, and wheat), but that this effect is statistically significant only for wheat. Finds no evidence that the effects of education should be attributed to family background or to ability. Measures of farmer modernity and agricultural knowledge are not correlated with farm efficiency.

754. John, Martha Tyler and Floyd Idwal John. 1990. "A Research Model Applied to a Computer Project in Swaziland." Pp. 43-60 in *International Science and Technology: Philosophy, Theory and Policy*, edited by Mekki Mtewa. New York: St. Martin's Press.

Examines the effectiveness of a computer training program (LOGO) initiated at a secondary school in Swaziland. Students produce more ideas, as measured by an Ideational Fluency Test, than students from other schools who have not had instruction. They also produce a wider range of ideas and show more positive attitudes toward computers, mathematics and languages.

755. Kahane, Reuven. 1976. "Education Towards Mediatory Roles: An Interpretation of the Higher Educational Policy in India in the Twentieth Century." *Development and Change* 7: 291-309.

Evaluates the role of higher education in mediating between core and periphery, modern and traditional approaches to social life.

756. Kahn, Michael J. 1990. "Access to Science Education in a Developing Country: The Challenge of Scientific Literacy." *Compare: A Journal of Comparative Education* 20: 155-162.

Briefly outlines formal changes in Botswana's scientific education programs since 1966, then disputes suggestions that Botswana's government has not been serious about scientific education. Attributes educational difficulties to implementation barriers rather than lack of will.

757. Kannan, K.P. 1981. "A People's Science Movement." *Development*. Pp. 37-40.

Summary of a report Kannan issued on a new group to develop rural science--The People's Science Movement in Kerala, a voluntary group started in 1962 for the popularization of science through the publication of popular books and magazines.

758. Kenney, Martin. 1987. "The University in the Information Age: Biotechnology and the Less Developed Countries." ***Development*** **4: 60-7.**

Examines the usefulness of biotechnology to these countries, taking into consideration socio-political aspects and distribution of gains from it.

759. King, Kenneth. 1986. "Mapping the Environment of Science in India." ***Studies in Science Education*** **13: 53-69.**

A general description of science in India, with particular attention to the popularization of science and to the role of science in the schools. Concludes: (1) more case study material needs to be made available, (2) priority should be given to building theory from Third World micro-initiatives, in order to see one of the major sources of criticism of current development theory, (3) more accounts of culture-dependent S&T are needed to counter the idea of a universal "culture-free" science, (4) funding agencies prepared to support research and dissemination of alternative knowledge frameworks should make more South-South learning possible, (5) materials for teaching alternative paradigms in schools need to be made available, and (6) ways of exposing policy research personnel in the various government ministries to alternative conceptions of skill, S&T, and agricultural extension should be explored.

760. Krugly-Smolska, Eva T. 1990. "Scientific Literacy in Developed and Developing Countries." ***International Journal of Science Education*** **12: 473-80.**

Discusses the meaning of scientific literacy, which is similar in developed and developing countries. Examines impediments to achieving scientific literacy. Suggests that science should be taught as a way of knowing with practical consequences. Describes recent (mid to late 80s) literature on scientific literacy.

761. Kwong, J. 1987. "In Pursuit of Efficiency: Scientific Management in Chinese Higher Education." ***Modern China*** **13: 226-56.**

Concentrates on the restructuring of education management to see how the Western concept of scientific administration has shaped academic administration in universities.

762. Lien, Da-Hsiang Donald. 1987. "Economic Analysis of Brain Drain." *Journal of Development Economics* 25: 33-44.

Analyzes factors leading students to foreign universities, and also offers a model to predict what proportion return home. The brain drain problem arises from the possibility of "signaling" from two-stage decision procedures within an asymmetric information framework. Signaling is defined as inference about the ability of individuals on the part of source countries from attributes of those individuals, which serve as quality signals. In two-stage decision procedures, individuals consider explicitly whether to return to the home country or stay abroad once they have already decided to go abroad. Causes of asymmetric information are attributed to institutional characteristics. Mathematical analysis.

763. Lien, Da-Hsiang Donald. 1988. "Appropriate Scientific Research and Brain Drain: A Simple Model." *Journal of Development Economics* 29: 77-88.

Decomposes research on the brain drain into a high-income characteristic and a low-income characteristic. Assuming individuals act so as to maximize their expected income levels, an upward bias in the social value of high-income type research creates a loss of allocative efficiency in human resource which parallels the case of an inappropriate product. A developing country should therefore not impart by policy an upward bias to high-income type research.

764. Lomnitz, Larissa. 1983. "Recruiting Technical Elites: Mexico's Veterinarians." *Human Organization* 42: 23-9.

Survey data from the School of Veterinary Medicine of the National University of Mexico used to describe how the technical-political leadership elite is recruited in the modern state of Mexico. Illustrated with four case histories. Views the new technical-political elites as coming largely from the middle classes and connected to the political system through the university. Mexican culture is important, since the recruiting of leaders takes place in the characteristic context of the patron-client relationship.

765. Macphee, C.R. and M.K. Hassan. 1990. "Some Economic Determinants of Third-World Professional Immigration to the United States." *World Development* 18: 1111-8.

Tests determinants of immigration to the United States by Third World engineers, natural and social scientists, and physicians. Data is from the INS and the National Science Foundation. Eighteen countries of origin. Principal explanatory variables are income, real GDP growth, graduation in the United States, and study in each country of origin. Other explanatory variables are foreign student enrollment in the U.S., lagged immigration, total immigration from each country, and a binary variable accounting for restrictions on permanent visas in 1972-73.

Authors use a variant of the 1959 Arrow-Capron model of dynamic shortages, which attributes immigration to labor shortages in the country of destination and labor surpluses in the country of origin. The results support the thesis that labor shortages in the U.S. contribute to the flow of these professionals from the Third World. They also support A.K. Sen's 1973 finding that U.S. education has an important influence on immigration. Restrictions on visas in 1972-73 were found to have no effect on immigration flow, except possibly in the case of physicians.

766. Maier, Joseph and Richard W. Weatherhead (eds.). 1979. *The Latin American University*. Albuquerque: University of New Mexico Press.

Rather dated now, this volume contains essays on origin and philosophy of the Spanish American, Brazilian, and Latin American university types, the European background, university reform, students, and professors.

767. Matthews, M. 1986. "Current Educational Practices in the People's Republic of China." *Asian Affairs* 17: 277-87.

Author is an English expert in Soviet studies who visited China for two weeks and compiled information after visiting primary, secondary, and high schools. But the account remains superficial because "model" schools were visited.

768. Mazrui, Ali A. 1979. "Churches and Multinationals in the Spread of Modern Education: A Third World Perspective." *Third*

World Quarterly 1 (1): 30-49.

Well-written ethnography on the diffusion of educational norms from the West to the South.

769. Meek, V. Lynn. 1982. *The University of Papua New Guinea: A Case Study in the Sociology of Higher Education.* St. Lucia: University of Queensland Press.

A history of the function, structure, and character of new universities in newly independent nations.

770. Meske, W. and M.C. Fernandez de Alaiza. 1990. "Structure and Development of the Scientific and Technological Potential in the Republic of Cuba." *Scientometrics* 18: 137-56.

The authors discuss their experiences while gathering statistics on the scientific and technological potential of the Republic of Cuba. They conclude that the structure of Cuba's educational system must be altered to increase emphasis on technical and natural science disciplines. Cuba also should beef up its ability to train more qualified post-graduate scientists, as well as increase financial incentive on the demand side of scientific deployment.

771. Molina Chocano, Guillermo. 1979. "The Training Process and Research in Central America." *International Social Science Journal* 31: 70-8.

Examines the teaching of social science in Central America. Looks in particular at the Central American School of Sociology.

772. Moravcsik, Michael J. 1975. *Science Development: The Building of Science in Less Developed Countries.* Bloomington, IN: International Development Research Center (2nd edition, 1976, PASITAM).

Stresses the need for domestic scientific education in developing countries, the termination of "brain drain," an improved scientific communication system within developing countries, the elimination of the international scientific communication system's bias against developing countries, and the need for both merit-based and institutional grants. Examines general problems common to all developing countries rather than specific problems by nation.

773. Morgan, W.R. and J.M. Armer. 1988. "Islamic and Western Educational Accommodation in a West African Society: A Cohort-Comparison Analysis." *American Sociological Review* 53: 634-9.

Tests the hypothesis that indigenous education systems are being replaced by Western systems of mass education. Using two surveys of youth in the city of Kano in northern Nigeria, conducted nine and fifteen years after independence, the authors find that attendance in both systems grew and increasingly accommodated each other. Nine-variable model of educational attainment is offered.

774. Munoz, Heraldo. 1980. "Social Science in Chile: The Institute of International Studies of the University of Chile." *Latin American Research Review* 15 (3): 186-9.

Description of the Institute of International Studies of the University of Chile, which was founded in 1966 with the cooperation of London's Royal Institute of International Affairs. A fundamental characteristic of the Institute is its Latin Americanist orientation. At the initiative of the director of the Institute, it organized and sponsored a number of major conferences, including one on "Science and Technology in the Pacific Basin."

775. Murphy, Terence. 1981. "Aspects of High-Level Manpower Forecasting and University Development in Papua New Guinea." *Journal of Developing Areas* 15: 417-34.

Analyzes the relationship between the potential for university advancement and the likely population at any given time, including an argument for universities in the Third World to allocate funds on a more utilitarian model.

776. Najafizadeh, Mehrangiz and Lewis A. Mennerick. 1990. "Educational Ideologies and Technical Development in the Third World." Pp. 29-42 in *International Science and Technology: Philosophy, Theory and Policy*, edited by Mekki Mtewa. New York: St. Martin's Press.

Authors suggest that Third World countries have been greatly influenced by Western educational ideology, which assumes that Western socio-political values and Western modernization and technical development are appropriate for all Third World countries. The authors feel that attention should be given to the possibility that alternative goals and methods may be more appropriate to some Third World countries. Concludes that Third World educational policy should attend to at least four major issues: (1) how Third World citizens see their educational needs, rather than how those needs are seen by governmental agencies, (2) how indigenous forms of education and indigenous values might be adapted to future education, (3) how incentives inhibiting educational change can be countered, and (4) an increased emphasis on technical training to facilitate development and adaptation of technologies.

777. Najafizadeh, Mehrangiz and Lewis A. Mennerick. 1988. "Worldwide Educational Expansion from 1950 to 1980: The Failure of the Expansion of Schooling in the Developing Countries." *Journal of Developing Areas* 22: 333-58.

Authors argue that the expansion of educational opportunities in the Third World has been overestimated by scholars, who glaze over differences in participation across nations and by gender. Furthermore, aggregate educational statistics hide disparities between primary and secondary school enrollments.

778. Odedra, Mayuri and Stefano Kluzer. 1988. "Bibliography for Information Technology in Developing Countries." *Information Technology for Development* 3: 297-356.

Highly useful source for the field of information technology.

779. Psacharopoulos, G. 1991. "Higher Education in Developing Countries." *Higher Education* 21: 3-10.

Special issue on higher education in developing countries, with an emphasis on strategies for financing in specific countries.

780. Radwan, Ann B. 1987. "Research and Teaching in the Middle East." *Annals* 491: 126-33.

Finds that the "professional quality and commitment" of the guest lecturer is the determining factor in successful individual experience and mutual understanding through educational exchange.

781. Ransom, Baldwin. 1988. "Education for Modernization: Meritocratic Myths in China, Mexico, the United States and

Japan." *Journal of Economic Issues* 22: 747-62.

How the levels of education in each of these nations attempt to preserve a meritocratic myth. Outline of educational organizations arguing that meritocracy hinders progress in both industrializing and developed countries.

782. Rao, K.N. 1978. "University-Based Science and Technology for Development: New Patterns of International Aid." *Impact of Science on Society* 28: 117-25.

The university is only one component of the national capacity in science and technology. Ideally universities work in concert with other institutions. Despite deficiencies, LDC universities have (1) responded to critical manpower needs (except in Africa and the Middle East) by producing large numbers of engineers, teachers, and scientists to staff development projects; (2) stimulated fellowship and exchange programs by international agencies, resulting in a rapid buildup of scientific cadres in many countries; (3) a mixed record of subprofessional technical training (training of technical teachers for polytechnics, junior colleges, and technical institutes); (4) a modest record of research.

Agricultural research tends to be directed at local problems but industrial R&D is still in its infancy, and scientific research tends to follow international currents. With basic scientific isolated from national needs, a network of scientists working on "common problems" hasn't been able to develop. Major issues: system planning at institutional level, issues of institutional efficiency, curriculum problems, and the need for outreach. For the purpose of international development assistance, countries are divided into three categories, depending on level of development of universities.

783. Saddique, Abu Bakr. 1989. "Acquisition of Scientific Literature in Developing Countries. 1: Bangladesh." *Information Development* 5: 15-22.

First of a series of articles in this issue on the acquisition of scientific literature by libraries in developing countries. Summarizes the development of science and technology, and scientific and technological libraries, in Bangladesh. Makes proposals for future action.

784. Sathyamurthy, T.V. 1984. "Development Research and the

Social Sciences in India." *International Social Science Journal* 36: 672-98.

The last two decades have seen an explosion of institutions of higher learning and research throughout India, but there is a lack of knowledge of each other's work at the institutional level. In no other developing country was such a pool of trained social sciences personnel available at the time of independence. Tries to set the emergence of India's fifty-odd institutes of research and development in a context of the development of ideas concerning social science research and its priorities as a part of independent India's intellectual history. Good historical schema, in stages, of the social sciences in India.

785. Schwartzman, Simon. 1986. "Coming Full Circle: A Reappraisal of University Research in Latin America." *Minerva* 24: 456-75.

A critique of reform movements aimed at developing Third World science. These movements ultimately decrease the autonomy of research and development institutions with what passes for "planning." The problem is serious because of the increasing importance of high technologies (electronics, biotechnology) that require institutional flexibility. Universities are a critical development asset.

786. Schwartzman, Simon. 1991. *A Space for Science: The Development of the Scientific Community in Brazil*. University Park, PA: Pennsylvania State University Press.

An important work on the history and development of Brazilian science. Based on published historical materials regarding Brazilian natural sciences, as well as interviews with important figures in the mid 1970s. General theme is the tension between a pragmatic view of science in terms of its economic and technological effects and an idealistic view of science as the search for new knowledge by a community of scholars.

Part Onc bcgins with an account of the intellectual heritage of Brazil in eighteenth century Portugal, the 19th century gold boom, and the establishment of the first professional schools and scientific groups under imperial sponsorship. The transition from imperial science to applied research in the early 20th century coincided with an influx of new immigrants and the emerging importance of Sao Paulo as an urban center. Centralization of power and a national concern with industrialization led to the founding of national universities. Part Two is an overview of the professionalization of Brazilian scientists in the twentieth century, generational differences, and the introduction of a modern scientific ethos. The final chapter recounts growth in the 1970s (especially in nuclear energy and computers) and decline in the 1980s.

787. Selvaratnam, Viswanathan. 1988. "Higher Education Co-Operation and Western Dominance of Knowledge Creation and Flows in Third World Countries." *Higher Education* 17: 41-68.

Outlines the "Eurocentric" elements in Western university knowledge systems, then points to productive and counterproductive effects of these features in Third World investments and research institutions. Discusses the possibility of South-South educational interchange for the development of a "self-reliant higher education system among Third World nation states," paying particular attention to information technology differences in informational resources.

788. Sharafuddin, A.M. 1986. "Science Popularization: A View from the Third World." *Impact of Science on Society* 36: 347-53.

Simplistic historical discussion of "how the Third World fell behind." Considers communication to laymen of simple scientific concepts, relating these to everyday life. Need to enlist local culture and use local languages in science popularization.

789. Shirk, Susan L. 1982. *Competitive Comrades: Career Incentives and Student Strategies in China*. Berkeley: University of California Press.

Written by a political scientist, this study of a city high school shows that intense individual competition (academic and political), rather than cooperation, is an unintended consequence of the revolutionary regime's attempt to create a "virtuocracy" (rewarding people who display the moral values of the regime). The consequences of allowing advancement as a function of "activism" include acrimonious political competition, avoidance of activists, retreat into the private worlds of family and friendship, and disaffection from regime.

790. Toh, Swee-Hin. 1977. "Canada's Gain from Third World Brain Drain, 1962-1974." *Studies in Comparative International Development* 12(3) : 25-45.

Uses multiple measures of the extent to which Canada has gained from Third World brain drain--all suggesting that Canada profits greatly by adding to its professional/technical stock without paying out the investment necessary to reap such social returns to education. The contribution by immigrants to net national income levels is found to be quite high, a factor that the author suggests is seldom considered when foreign aid is appropriated.

791. Toure, Saliou. 1988. "The Promotion of Science in an African Country." ***Impact of Science on Society*** **152 : 363-71.**

Discusses a campaign of science popularization in the Ivory Coast carried out with involvement of government ministries and agencies. Examples are publications, radio broadcasts, and TV programmes of the Ministry of Agriculture, the Ministry of Scientific Research, and the Ministry of Primary Education.

792. Ukaegbu, Chikwendu C. 1985. "Educational Experiences of Nigerian Scientists and Engineers: Problems of Technological Skill-Formation for National Self-Reliance." ***Comparative Education*** **21: 173-82.**

Summarizes highlights from a study on indigenous high-level scientific and technological manpower in Nigeria. Finds that positive educational experience is higher among foreign-trained than locally trained Nigerian scientists and engineers.

793. Ukaegbu, Chikwendu C. 1985. "Are Nigerian Scientists and Engineers Effectively Utilized? Issues on the Deployment of Scientific and Technological Labor for National Development." ***World Development*** **13: 499-512.**

Suggests the qualitative manner in which scientists are used poses at least as big an obstacle to development as the quantity of scientists, because labor isn't fully used. Survey of 266 scientists and engineers.

794. Van den Bor, W. and J.C.M. Shute. 1991. "Higher Education in the Third World: Status Symbol or Instrument for Development?" ***Higher Education*** **22: 1-16.**

Presents an analysis of problems originating from national policy conditions and institutional weaknesses that prevent development of Third World higher education despite heavy investment. This is done through a critical analysis of the World Bank report on "Education in Sub-Saharan Africa: Policies for Adjustment, Revitalization, and Expansion." Also describes the results of a recently published comparative study of higher agricultural institutions in 10 countries in Asia, Africa, and Latin America. Ends by suggesting ways to improve higher education in the South by using South-North university cooperation.

795. Velho, Lea. 1990. "Sources of Influence of Problem Choice in Brazilian University Agricultural Science." ***Social Studies of Science*** **20: 503-17.**

Brazilian scientists tend to choose research topics that they perceive are directly relevant to local problems, straying from lines of agricultural research established abroad. Four universities, scientists in two subfields studied.

796. Weis, Lois. 1981. "The Reproduction of Social Inequality: Closure in the Ghanaian University." ***Journal of Developing Areas*** **16: 17-30.**

Examines whether university funding in Third World serves to maintain existing pattern of structured social inequality. Looks at student backgrounds and finds decreasing fluidity of recruitment between 60s and 70s.

Chapter 10 -- Distribution

797. Adas, Michael. 1989. ***Machines as the Measure of Man: Science, Technology, and Ideologies of Western Dominance.*** **Ithaca: Cornell University Press.**

Historical study of the role of science and technology in shaping the European response to non-Western peoples. Argues that imperialist attitudes toward non-Westerners were rooted in European mastery of the physical world through technique and suggests that World War I shook Europe's confidence in itself and in the unquestionable value of technique. After World War II, the modernization paradigm replaced the ideology of the civilizing mission, retaining technology as a standard of superiority, but repudiating racism and placing greater emphasis on political aspects of development.

798. Ahmed, Iftikhar. 1986. "Technology, Production Linkages and Women's Employment in South Asia." ***International Labour Review*** **126: 21-40.**

Looks at the social impact, for women, of technological change in South Asia. Concludes that modernization has done little to free women from their traditional domestic roles. The references will be very useful for those interested in economic or technical issues regarding women in South Asia.

799. Alavi, Reza. 1985. "Science and Society in Persian Civilization." ***Knowledge: Creation, Diffusion and Utilization*** **6: 307-28.**

This issue of the journal examines the historical experience of science development in China, India, Iran, and Japan. Of these four, the Muslim world has remained most impervious to modern sciences. Gives the historical background of science in Persia/Iran. In modern Persia the issue has been one of intelligentsia versus clergy. Two nations developed from 1953-1980: a ruling minority with a cultural orientation alien to the majority, and a nation of marginals. The economy of marginals is relatively autonomous.

800. Barnes, Douglas F. 1988. *Electric Power for Rural Growth: How Electricity Affects Rural Life in Developing Countries.* Boulder: Westview Press.

Defines and addresses a broad set of issues on the efficacy of rural electrification for development. Seven research questions: (1) Does rural electrification generate additional productivity, higher income, and structural change in rural areas? (2) How does it fit into a broad strategy of rural development? (3) What are the effects on equity? (4) What are the benefit-cost ratios in financial and social terms? (5) What are the most effective types of electrification strategies? (6) How does central grid electrical service compare to decentralized forms of electricity and energy production? (7) How do different electrification policies affect the impact and implementation of electrification?

801. Belote, James and Linda S. Belote. 1984. "Suffer the Little Children: Death, Autonomy and Responsibility in a Changing 'Low Technology' Environment." *Science, Technology, and Human Values* 9(4): 35-48.

Discusses the likely impact of technological change on family values in Ecuador. Saraguro parents once raised their children in a way that promoted autonomy and fit with the high infant mortality rates expected in this society. However, with increasing technological complexity and a rising cost of rearing children, the parents are found to be developing greater "engagement" with their offspring.

802. Chadney, James G. 1984. "The Economic Implications of the New Technologies in Punjab." *Eastern Anthropologist* 37: 227-37.

An attempt to document and analyze some of the socio-economic changes that have occurred as a result of the Green Revolution in India, using data from Punjab. Concludes that an obvious disparity exists between developed and less developed villages and between owner/cultivators and agricultural laborers.

803. Critchfield, Richard. 1982. "Science and the Village: The Lost Sleeper Awakes." *Foreign Affairs* 61: 14-41.

The 1970s were a turning point in the Third World similar to the 1920s and 1930s in American agriculture. The change came when villagers saw concrete evidence of how Western tech could improve lives. But their Agricultural Revolution differs from ours owing to large populations, its occurrence 50 years later with more advanced technology, and a peasant culture highly dependent upon the village. Describes six main cultural variations of the Third World--Confucian, Malay-Javanese, Hindu, Xian, Islamic, and African--to explain why villagers adjust differently to Western ideas and technology. American experience and advice valuable to avoid rural depopulation. Priority should also be given to the transfer of sophisticated Western technology and ways to harness old resources. The author is wary of AT emphasis on energy technology.

804. Date-Bah, Eugenia and Yvette Stevens. 1981. "Rural Women in Africa and Technological Change: Some Issues." *Labour and Society* 6: 149-62.

Probes the disturbing finding that development in Third World rural areas does not reduce the workload for females, and if anything increases it. Emphasizes the place of rural women in the Third World family unit, culturally imposed gender roles, the heterogeneity of women in various LDC agricultural sectors, existing traditional groups, and marketing barriers for innovations--and the problems each of these pose for aiding Third World rural women.

805. Dillman, C. Daniel. 1983. "Assembly Industries in Mexico: Contexts of Development." *Journal of Interamerican Studies and World Affairs* 25: 31-58.

Study of the maquiladora industries that have arisen in Mexico since the 1960s. Finds that Mexico's new administrative procedures, employment practices, and fiscal incentives, plus devaluations of the peso, led to resurgent production from existing plants and construction of new facilities. However, problems arose, since the maquiladoras were industrial enclaves that failed to reduce Mexican unemployment and underemployment, tended to pull migrants from the interior to the border cities, and increased dependence of the border zone on the United States. The low level of domestic inputs hindered backward linkages and discouraged growth of ancillary industries.

806. Edquist, Charles. 1985. *Capitalism, Socialism and Technology: A Comparative Study of Cuba and Jamaica*. London: Zed Books.

Compares socio-economic aspects of technical change in sugar-cane harvesting in Cuba and Jamaica. In the late 1950s both countries were capitalist. By the early 1980s, Jamaica was capitalist and Cuba was socialist. This makes it possible to compare changes in the two (the author minimizes the socialist phase in Jamaica and the role of the Soviet Union in Cuban economy). Suggests that socialist Third World countries may have an advantage over capitalist ones in the introduction of advanced techniques since they have more efficient socio-economic and political mechanisms to deal with displaced workers. Cost-benefit analyses were insufficient as a basis for decision making with regard to mechanization in both countries.

807. Foster, G.M. 1973. *Traditional Societies and Technological Change* (2nd edition). New York: Harper and Row.

Update of volume originally published in 1962, with anthropological focus. The concept of "limited good" explains traditional peasant behavior. Urbanization and nationalism are emphasized in a framework involving cultural, social, and psychological barriers to change. Chapters on technical specialists (ego gratification needs are discussed) and bureaucrats, and the potential contribution of anthropologists.

808. Garmany, J.W. 1978. "Technology and Employment in Developing Countries." *Journal of Modern African Studies* 16: 549-64.

Discusses issues of technology choice in relation to providing employment. Fails to reach definite conclusions and calls for empirical investigations and awareness of the problem. Few references.

809. Gutierrez, Fernando Calderon. 1988. "Crisis, New Technologies and Social Movements." Pp. 259-275 in *Science, Technology, and Development*, edited by Atul Wad. Boulder: Westview Press.

Considers the economic crisis of the late 80s and what role new technologies will play in new social formations, given this crisis. Sees social movements, such as feminism, ecology, and human rights as "dispersed resistance." Suggests that power will be further concentrated in the State. Obscure language.

810. Hainsworth, Geoffrey B. 1982. "Beyond Dualism." In

***Village-level Modernization in South East Asia*, edited by Geoffrey B. Hainsworth. Vancouver: Univ. of British Columbia Press.**

Argues that agricultural innovation and improvement is a continual process in most traditional agrarian societies. The language is excruciating: "There is a marked difference, however, between this gradual or occasional and substantially villager-controlled change, introduced alongside or as an extension of customary practices and generally reversible or amendable and the abrupt, pre-packaged, all-pervasive, and largely exogenously-forced or promoted change incurred with the Green Revolution and its concommittants [sic]."

811. Hammoud, H.R. 1986. "The Impact of Technology on Social Welfare in Kuwait." *Social Service Review* 60: 52-69.

Investigates the impact of technology on the modernization process of Kuwait's oil-rich society, finding both positive and negative effects on the Arab country's emerging social welfare system. Looks in particular at education, health, and housing, but suggests the generosity of government programs is producing growing dependency and rising expectations, a concern since the nation's welfare state relies so heavily on fluctuating revenues brought by petroleum.

812. Hanson, Jarice and Uma Narula. 1990. *New Communication Technologies in Developing Countries*. Hillsdale, NJ: Lawrence Erlbaum Associates.

Uses a social context approach to consider the effects of communication technology on developing countries. Focuses on "social technologies," defined as ways of doing things that contribute to changes in social systems. Social technology is said to be critical to technological change. The first three chapters establish the theoretical approach, by introducing the concept of social technology, examining notions of the information society, and reviewing main theories and applications of information technology.

The empirical sections consist of case studies of the information technology model in India, in other nations in South Asia and in ASEAN and the Pacific communities, in Latin America, and in the Arab nations. Brief conclusions regarding applications to policy decisions, the social technology perspective, and the information society. Makes recommendations on appropriate technology (referred to as "cost access"), cost effectiveness, and the possession of technology.

813. Hess, David J. 1991. *Spirits and Scientists: Ideology, Spiritism, and Brazilian Culture*. University Park, PA: University of Pennsylvania Press.

Discusses Brazilian Spiritism (otherwise known as *kardecismo*, after 19th-century French educator Allan Kardec). Considers Spiritist intellectuals as mediators between popular religion (such as Umbanda) and "legitimate" forms of knowledge recognized by universities, the state, and the medical profession. Discusses the role of Spiritism in religion, science, political ideology, medicine, and the social sciences. Hess's examination of the Spiritist psychotherapy of Menezes (set forth in Menezes' book, *Insanity Through a New Prism*) is particularly interesting.

814. Hill, Jill. 1990. "The Telecommunication Rich and Poor." *Third World Quarterly* 12(2): 71-90.

Attempts to explain how the gap between the telecommunications rich and poor has widened over the past twenty years, how the ideological framework has altered, and what steps developing countries have taken and are taking to respond to the pressures and constraints. Among other points, maintains that a decrease in multilateral funding has promoted the rise of the World Bank and the International Telecommunications Union (ITU) as mediators between MNCs and developing countries, and that the market restructuring urged on developing countries has increased the separation of telecommunications from government, and that restructuring (liberalization) may have unacceptable distributional consequences and political risks.

815. Hoffman, Kurt. 1985. "Clothing, Chips, and Competitive Advantage: The Impact of Microelectronics on Trade and Production in the Garment Industry." *World Development* 13.

Shows how the introduction of microelectronics technology was limited, maintaining most labor-intensive work patterns. Also looks at R&D differences in the garment industry.

816. Hoshino, Shinya. 1982. "Problems of Human Adaptation in the Context of Changing Patterns of Technology." *Indian Journal*

of Social Work **42: 352-63.**

Very broad treatment of adaptational problems posed by changing technology, with special attention to environmental problems, problems of distribution, and problems posed by rising populations. Suggests that government programs can only reduce fertility if lower fertility can be made a rational choice for most of the population.

817. Jaireth, Jasveen. 1988. "Class Relations and Technology Use: A Study of Tubewell Utilization in Punjab (India)." ***Development and Change*** **19: 89-114.**

Studies the utilization of tubewells by different socio-economic categories of users in Punjab. First discusses modes of tubewell irrigation currently prevalent in Punjab and discusses implications for cost and efficiency of irrigation. Second part looks at the distribution of modes of irrigation over different economic categories of cultivators. Finds a concentration of poorer modes of irrigation (in both cost and efficiency) among smaller cultivators.

818. James, Jeffrey and Frances Stewart. 1982. "New Products: A Discussion of the Welfare Effects of the Introduction of New Products in Developing Countries." Pp. 225-55 in ***The Economics of New Technology in Developing Countries*****, edited by Frances Stewart and Jeffrey James. Boulder, CO: Westview.**

New products are often inegalitarian in impact and can lead to welfare losses. Eight cases examined in which results are mixed, but with many undesirable side effects.

819. Junker, Louis. 1983. "The Conflict Between the Scientific-Technological Process and Malignant Ceremonialism." ***American Journal of Economics and Sociology*** **42: 341-52.**

A technological revolution may be seen as the process in which more flexible or warranted technological relations break through destructive forces so decisively that the institutional and technological structure is transformed into efficient organizational structures (Clarence Ayres). Two processes are at loggerheads: encapsulation versus liberation. A general trust in technology abandons the main thrust of the ceremonial-instrumental dichotomy: the scientific-technological process is the social context in which the forces of warranted or warrantable knowledge are expressed through institutions enlarging accessibility and participation on a peer-to-peer basis.

820. Kenney, Martin. 1983. "Is Biotechnology a Blessing for Less Developed Nations?" *Monthly Review* 34(11): 10-19.

Outlines the trajectory of biotechnological development and use, describing how, under a capitalist world economic system, the end result of this "miracle" technology will result in increased hardship and exploitation in the Third World. Offers strategies for developing countries to harness biotechnology despite transnational capitalist production advantages.

821. Lent, John A. 1991. "The North American Wave: Communication Technology in the Caribbean." Pp. 66-102 in *Transnational Communications: Wiring the Third World*, edited by Gerald Sussman and John A. Lent. Newbury Park, CA: Sage.

Examines how dependency in the field of communication technology occurred in the Caribbean and how this has affected local economies and cultures. Information on North American and transnational penetration of the Caribbean focuses on the 1980s. Cuba offers the only successful alternative to North American domination.

822. Malik, Yogendra K. 1982. "Attitudinal and Political Implications of Diffusion of Technology: The Case of North Indian Youth." *Journal of Asian and African Studies* 17: 1-12.

Studies the interrelationship between specific components of modern technology and the development of attitudinal patterns in a selected group of individuals--focuses on the development of (1) personal competence (demonstrated by achievement orientation), (2) its relationship to the development of democratic dispositions, (3) growing sense of party identification, and (4) political alienation. Employs a Guttman scale of exposure to technology and achievement orientation.

Empirical tests carried out in Jullandur City of Punjab. Exposure to tech is not related to alienation, but alienation is related to discontinuities and tensions caused by cultural changes. Higher levels of exposure to technology leads to development of attitudinal profile supportive of democracy. There are penetrating and emulative linkages of dominant and dependent cultures.

823. Melody, William H. 1991. "The Information Society: The Transnational Economic Context and Its Implications." Pp. 27-41 in *Transnational Communications: Wiring the Third World*, edited by Gerald Sussman and John A. Lent. Newbury Park, CA: Sage.

Examines the implications of changes in information and communication for developing country institutions, development, and policy. With regard to effects on economic efficiency and market extension, benefits of the new technologies will be unevenly distributed, increasing relative and absolute poverty of the less-advantaged segments of society. Competitive advantage of products in the market is obtained primarily by persuasion of government leaders, rather than by product superiority. With regard to the role of national governments, TNCs are becoming more direct instruments of macroeconomic policy for developed countries; national governments that tie themselves closely to TNC interests are limiting their own freedom of action.

Traditional market theory, which assumes that technology is autonomous and beneficial, states that oligopolistic competition in foreign markets is free-market competition, and that optimum resource allocation can be obtained by seeking short-run profits. This provides a rationale for expansion of TNC power. But the model of a national public service monopoly for communications systems is even worse than private monopoly. Oligopolistic rivalry among TNCs creates an opportunity for Third World countries to develop communication systems that serve local needs.

824. Morehouse, Wade (ed.). 1988. *Science, Technology and the Social Order*. New Brunswick, New Jersey: Transaction Books.

Reprinted from a special issue of *Alternatives: A Journal of World Policy* (1978-79, volume 6), for which Morehouse served as guest editor. Papers are generally critical of capitalism and of developed countries. Appendices contain two documents. The first, "The Perversion of Science and Technology: An Indictment," is a statement condemning "... the way in which science and technology have become instruments of a global structure of inequality, exploitation, and oppression," signed by 26 individuals at the 14th meeting of World Order Models Project in Poona, India, July 2-10, 1978. The second document is the "Pugwash Guidelines for International Scientific Cooperation for Development," drafted at an international Pugwash Workshop in Haryana, India, January 11-14,

1978.

825. Muga, D.A. 1987. "The Effect of Technology on an Indigenous People: The Case of the Norwegian Sami." ***Journal of Ethnic Studies*** **14: 1-24.**

Study of the effect of technology on a Norwegian minority group. Focuses on the use of the snowmobile in reindeer herding. This technology has had a positive effect on the Sami by contributing to the socialization of the production process, which can lead to a socialist economy.

826. Najafizadeh, Mehrangiz and Lewis Mennerick. 1989. "The Impact of Science and Technology on Third World Development: Issues of Social Responsibility." ***Social Development Issues*** **12: 1-10.**

Considers ways in which technology may serve constructive or destructive ends for Third World countries. A broad, simplified introduction to issues of the social, political, and economic effects of technology.

827. Nandy, Ashis. 1979. "The Traditions of Technology." Pp. 371-385 in ***Science, Technology and the Social Order,*** **edited by Ward Morehouse. New Brunswick, NJ: Transaction Books.**

Argues that modern technology is a particular form of traditional technology that has developed over the past 300 years. It has become the dominant technology by marginalizing other technologies in the West and non-Western worlds. The author finds modern technology to be alienating, exploitative, and dehumanizing. To correct this, an alternative ideology of science and a new legitimacy for traditional technosystems and their cultural environments is necessary.

828. Nichols, Grace Olney. 1977. "Hydroelectric Development in Guyana." ***Impact of Science on Society*** **27: 321-30.**

Describes the Upper Mazaruni hydroelectric scheme and the organization charged with the project, the Upper Marazuni Development Authority. Also discusses problematic changes this will bring to the lives of two Amerindian tribes whose lands will be inundated as a result.

829. Oliveira, Omar Souki. 1991. "Mass Media, Culture, and

Communication in Brazil: The Heritage of Dependency." Pp. 200-213 in *Transnational Communications: Wiring the Third World*, edited by Gerald Sussman and John A. Lent. Newbury Park, CA: Sage.

Maintains that the mass media in Brazil foster a culture of consumerism that perpetuates economic inequality and dependency. Television advertising, in particular, is said to encourage consumption of Western products and discourage consumption of indigenous products. This contributes to greater income concentration.

830. Pickett, James and R. Robson. 1977. "Technology and Employment in the Production of Cotton Cloth." *World Development* 5: 203-15.

Analyzes the effect of technology on employment and production, using two UN studies on the spinning and weaving of cotton cloth. Finds that (1) a business would have to spend immense amounts of money to increase technological efficiency without reducing employment and (2) increased efficiency in developing textile industries could cause increased unemployment.

831. Pool, Ithiel de Sola. 1990. *Technology Without Boundaries: On Telecommunications in a Global Age.* Cambridge and London: Harvard University Press.

Posthumously published work of a major political scientist and director of the MIT Research Program on Communications Policy. Comprehensive view of the social, political, and cultural effects of new communications technology. The book is divided into three parts: (1) Communications and the Changing Environment, (2) Satellites, Computers, and Global Relations, and (3) Ecology, Culture, and Communications Technology.

A major theme is the capacity of the new technology to break down political and geographic boundaries. The author is critical of media policy, which he sees as repressing diversity and enforcing centralization. The second chapter of Part I offers an excellent non-technical description of the new technologies (which the author sees as consisting of 25 main devices), a description of digital versus analog signals, a discussion of the different types of transmission media, an explanation of spectrum, a brief overview of communications satellites, and a review of computer message processing.

832. Rao, K. Nagaraja and Joel B. DuBow. 1984. "The Allure of Optimum Technologies and the Social Realities of the Developing World." ***Bulletin of Science, Technology and Society*** **4: 345-55.**

Although experts have begun to realize the need for selecting an optimal technology geared to specific development situations, this push for new technological capabilities still overlooks concerns about successful implementation and doesn't take into account certain "social realities"--the viewpoints, objectives, and constraints for the targeted group. Offers an approach for defining a situation's "optimum technology" and illustrates with a small-scale case study on rural hydroelectric systems in India.

833. Richter, Maurice N., Jr. 1982. ***Technology and Social Complexity*****. Albany: State University of New York Press.**

An informative theoretical examination of the relationship between technological innovation and social change. Takes an evolutionary approach to society and sees evolutionary concepts as more fundamental than developmental concepts. Identifies technological innovation as a decisively important source of social-evolutionary change. Suggests that we have two general weaknesses in our knowledge: (1) different perspectives and research traditions are only imperfectly integrated; (2) studies of social conditions that make technological innovation possible and studies of social impacts of technology have yielded considerable information on particular types of situations, but not much information to lead to general knowledge about conditions under which technological innovation is likely to occur.

834. Santiago, Carlos E. and Erik Thorbecke. 1984. "Regional and Technological Dualism: A Dual-Dual Development Framework Applied to Puerto Rico." ***Journal of Development Studies*** **20: 271-89.**

Shows how dual economies based both upon urban-rural and traditional-modern dichotomies exist at the same time. This is referred to as a "dual-dual" framework. The article applies this framework to Puerto Rico and finds that Puerto Rico's "industrialization first" development strategy resulted in (1) a concentration of income and employment in the urban sector, (2) the virtual disappearance of informal agriculture and a shrinking of the rural farm sector, and (3) a very limited role for the informal urban sector in absorbing labor.

835. Schumann, Gunda. 1984. "The Macro- and Microeconomic Social Impact of Advanced Computer Technology." ***Futures*** **16: 260-85.**

Investigates the impact of advanced computer technologies on relationships between headquarters and affiliates in management, employment, and work environment, as well as the economic role of women and the economies of LDCs. Argues that computers and information technologies under the control of transnational corporations that try to gain competitive advantages tend to concentrate power at headquarters. They increase the rate of unemployment, alienate workers from their work environment, reinforce traditional economic and social roles of women, and widen the gap between developed countries and LDCs.

836. Shaw, Anthony B. 1984. "Impact of New Technology on the Guyanese Rice Industry: Efficiency and Equity Considerations." ***Journal of Developing Areas*** **18: 191-218.**

Addresses the efficiency and equity of newly introduced seed-fertilizer technology in the rice economy of Guyana. Addresses the questions: (1) Are there significant variations in the quantity of related factor input use among adopters of the new technology and nonadopters? (2) Do large farms use higher quantities of modern inputs than small farms? (3) To what extent have new varieties of rice and related inputs increased per-acre yield and income over traditional varieties? (4) Has the new technology changed the relationship between farm size and farm productivity?

Findings are based on microlevel data from the Essequibo Coast region through a survey of 125 farmers during the crop year 1979. Finds that actual per-acre income was lower for adopters than nonadopters because of the increase in number and costs of purchased inputs. However, large-sized and irrigated farms used more nontraditional inputs and showed higher per acre yields. The new technology therefore disproportionately benefits large farms. New government subsidization policies are suggested to correct this.

837. Slamecka, Vladimir. 1985. "Information Technology in the Third World." ***Journal of the American Society for Information Science*** **36: 178-83.**

Suggests that new Third World developments in information

technology have had generally positive effects in the social and cultural realms, while conceding the economic impact favors industrial countries rather than those of labor-intensive industries. Specifically looks at effects of information technology on national "informatics policies," suggesting the "postindustrial society" might not be a realistic goal of Third World social development.

838. Stamp, Patricia. 1989. *Technology, Gender, and Power in Africa*. Ottawa: International Development Research Centre.

Case studies from the development literature on agriculture, health, and nutrition, exploring the interactive relationship between technology transfer and gender issues. Examines different conceptual frameworks, identifies factors that render women powerless and disadvantaged. Empowerment of women at the local level is vital to development.

839. Stewart, Frances. 1978. "Inequality, Technology and Payment Systems." *World Development* 6: 275-94.

Considers the way technology, population growth and the payments system are responsible for growing inequality in poor countries, and looks at their interrelationships. Payment systems are defined as the rules governing property rights, access to work, and income from work. Main types are traditional, capitalist, mixed, and socialist. Argues that capitalist and mixed payment systems are responsible for situations in which poverty has increased despite growth in per capita income. Suggested reforms focus on redistribution of income.

840. Turnbull, David. 1989. "The Push for a Malaria Vaccine." *Social Studies of Science* 19: 283-300 .

Discusses socio-political obstacles to developing a vaccine in Papua New Guinea. Claims research is more likely to benefit tourists and the military than the island inhabitants. This is due to the factors that seem to make development of a vaccine inevitable, which include the importance of a laboratory-based approach to dealing with malaria, the economic circumstance of vaccine production, the problems concerning the variability and specificity of malaria, and the significance of the socio-political situation in Papua New Guinea for medical research.

841. United Nations Industrial Development Organization. 1984. "Technological Self-Reliance of the Developing Countries: Toward Operational Strategies." Pp. 97-174 in *Technology Policy and Development: A Third World Perspective*, edited by Pradip K. Ghosh. Westport, CT: Greenwood Press.

Argues that technology incorporates, reflects, and perpetuates value systems. Its transfer thus implies the transfer of structure.

842. Walsham, Geoff, Veronica Symons and Tim Waema. 1988. "Information Systems and Social Systems: Implications for Developing Countries." *Information Technology for Development* 3: 189-204.

Argues that information systems should not be seen as technical systems that involve social behavior but as social systems, with technology as just one of the elements in the system. The authors approach information systems using Peter Checkland's "soft systems" methodology and Kling's "web models." The first is an approach to problematic situations that involves seeing that purposeful activities can be described in different ways in different world views.

Conceptual models are compared to perceived reality in order to gain insight into the problems. The web model sees computer systems as a web of resources that are meaning-charged social objects as well as tools. In order to test the transferability to developing countries of their alternative approach the authors are currently designing a case study of a financial service sector organization in Kenya, using adaptations of methodologies used in the UK. A number of case studies in developing countries are discussed in which the development and use of information systems can be best understood by including analysis of social, organizational and political context.

843. Yuchtman-Yaar, E. and A. Gottlieb. 1985. "Technological Development and the Meaning of Work: A Cross-Cultural Perspective." *Human Relations* 38: 603-21.

Reports on perceptions of the extent of technological change and its subjective impact among workers in five industrialized nations. Finds that perceived technological change varies significantly among the sampled publics, probably due to both technological change itself and to socio-cultural differences. The implications of the findings are discussed.

844. Yun, Hing Ai. 1986. "Science, Technology and Development." ***Journal of Contemporary Asia*** **16: 144-80.**

Impact of science and technology on development will be limited as developing countries are still embedded in socioeconomic and political structures based on unequal relations between domestic groups and with advanced capitalist countries. Problems in Malaysian development are seen as involving technology transfer rather than science transfer. Opposes technological determinism, arguing that competing institutions and social groups determine the path of scientific and technological development.

In the early 1950s, the paradigm stressed by LDCs involved acquisition of scientific capability and creation of industrial base. Production R&D was unattractive; overemphasis on big science. Next emphasis: technology transfer and adaptation--but AT still focuses on gadgets and techniques rather than on the development of technological skills. Concludes that recent trend in Malaysia has been rapid acquisition of expensive and sophisticated technology.

Chapter 11 -- Environment and Health

845. Alvarez, Gonzalo. 1982. "The Neurology of Poverty." *Social Sciences and Medicine* 16: 945-50.

Discussion of actual physiological problems resulting from poverty that hamper education. Considers how this affects the Third World.

846. Banta, H. David. 1986. "Medical Technology and Developing Countries: The Case of Brazil." *International Journal of Health Services* 16: 363-73.

Shows how nations make the wrong choices of technologies, and suggests reasons: a lack of expertise for decision making, lack of indigenous technology, and lack of policy research to guide actions.

847. Bowers, John Z., J. William Hess, and Nathan Sivin. 1988. *Science and Medicine in 20th Century China: Research and Education.* Ann Arbor, MI: University of Michigan.

Essays covering both the Republican period and the modern People's Republic, with specific essays on genetics, taxonomy, geology, biomedical research, population policy, viral vaccines, agriculture, nutrition, and education.

848. Bowonder, B. 1987. "The Bhopal Accident." *Technological Forecasting and Social Change* 32: 169-82.

Examines the events preceding and following the release of methylisocyante at the Union Carbide plant in Bhopal. Finds that human, technological, and system errors occurred. Suggests changes in policies for toxic-materials handling in developing countries.

849. Dike, Azuka A. 1985. "Environmental Problems in Third World Cities: A Nigerian Example." *Current Anthropology* 26: 501-5.

Gives examples of environmental problems from the city of Onitsha, Nigeria, to demonstrate that environmental pollution is now obvious in developing countries. Data from an opinion survey of residents living in four different zones of the city. Survey questions concern types of toilets, water flow in the gutters, common illnesses suffered, perceived causes of traffic congestion and noise, the degree to which noise is considered disturbing to work concentration, and refuse disposal. The major visible form of environmental pollution is the dumped heaps of refuse in the streets of the city. Onitsha is said to be generally representative of Nigerian cities.

850. Dixon, Robert K. 1988. "Forest Biotechnology Opportunities in Developing Countries." ***Journal of Developing Areas*** **22: 207-18.**

Forest biotechnology is an important tool for improving the resource efficiency and sustainability of food, fuel, and fiber production in developing countries. An expanded commitment to public and private research at the national and international levels is needed to ensure that developing world priorities are addressed.

851. Flavin, Christopher. 1983. "Photovoltaics: A Solar Technology for Powering Tomorrow." ***Futurist*** **17: 41-50.**

Popular description of the advantages of photovoltaics and of the progress of the photovoltaic industry by a researcher with the Worldwatch Institute. Suggests that this can be an important source of energy for the Third World, even for most electrical needs in villages.

852. Hoffman, Kurt. 1980. "Alternative Energy Technology and Third World Rural Energy Needs: A Case of Emerging Technological Dependency." ***Development and Change*** **11: 335-66.**

The West will be able to develop and market alternative energy techniques before the Third World. This raises the danger that developing countries may find themselves again in a situation of technological dependence. Concludes that developing countries may have to develop more aggressive strategies for technology acquisition.

853. Hoffman, Kurt. 1985. "The Commercialization of Photovoltaics in the Third World: Unfulfilled Expectations and Limited Markets." ***Development and Change*** **16: 5-38.**

Investigation of hypotheses advanced in author's earlier work on photovoltaic energy conversion systems. Suggests that photovoltaic systems have many advantages but the chief reason they are in limited worldwide use is cost. Concludes that it will take longer than anticipated for demand to reach the take-off point. This suggests developing countries should avoid making any long-term commitment to extensive investment in photovoltaic systems.

854. Inhaber, Herbert. 1985. "Risk in Developing Countries." ***Risk Analysis*** **5: 87.**

Guest editorial regarding a 1984 plant explosion in Mexico City (resulting from a liquified gas processing plant owned by Pemex) and the Bhopal accident in India.

855. Jasanoff, Sheila. 1986. "Managing India's Environment." ***Environment*** **28(8): 12-16, 31-8.**

Account of India's increased attention of environmental issues in the wake of the Bhopal incident. Includes a discussion of obstacles to successful implementation of environmental laws, such as decentralization of environmental decision making and the lack of judicial expertise in environmental issues.

856. Johns, D.M. 1990. "The Relevance of Deep Ecology to the Third World: Some Preliminary Comments." ***Environmental Ethics*** **12: 233-52.**

The author responds to an article by Ramachandra Guha in *Environmental Ethics* 11 (1989):71-83 which argues that deep ecology is flawed because it falsely equates environmental protection with wilderness preservation. Johns counters that deep ecology's distinction between anthropocentrism and biocentrism is useful in dealing with two central global problems--overconsumption and militarism.

857. Katz, James Everett and Onkar S. Marwah. 1982. ***Nuclear Power in Developing Countries.*** **Lexington, MA: D.C. Heath.**

A collection of essays dominated by country studies--Argentina, Brazil, China, Egypt, India, Indonesia, Iran, Korea, Mexico, Pakistan, the Philippines, Taiwan, Turkey, Venezuela, Yugoslavia. Also surveys nuclear-energy programs, the role of science and government, and the economy of nuclear power.

858. Khanom, Kurshida and Robert C. Leonard. 1989. "A Hygiene Experiment in Rural Bangladesh." ***Sociological Perspectives*** **32: 245-55.**

Measures the success of hygiene knowledge diffusion, and tries to assess which barriers are most important. Based on a pre-test, post-test health education experiment conducted with 162 Muslim farming families in a single village. Finds that one month's intensive effort by 10 student health educators produced considerable change toward a more preventive orientations. Results differed according to social class, with the largest effect for the middle class.

859. Lovins, Amory B. 1977. *Soft Energy Paths*. New York: Harper & Row.

The dean of soft energy expresses it best: "The huge capital intensive energy facilities often proposed to end unemployment not only make it worse, by draining from the economy capital that could make more jobs if invested in almost anything else, but also worsen inflation by tying up billions of dollars nonproductively for a decade. And unemployment and inflation are only the first of a long list of distressing side effects of a high growth, high technology, high risk approach to our energy problems."

860. Luddemann, Margarete K. 1983. "Nuclear Power in Latin America: An Overview of Its Present Status." *Journal of Interamerican Studies and World Affairs* 25: 377-415.

Examines nuclear policy, power forecasts, cooperation efforts, and approaches to nonproliferation in Argentina, Brazil, Mexico, and Cuba. Short descriptions of nuclear research and construction of nuclear power plants in Chile, Colombia, and Venezuela. Concludes that mutual confidence in international nuclear relations seems paramount and that nuclear technology may have the effect of making its users more interdependent.

861. Martinez-Polomo, Adolfo. 1987. "Science for the Third World: An Inside View." *Perspectives in Biology and Medicine* 30: 546-57.

Attributes the difficulties of Third World countries with scientific/technological development to the widespread poverty found there, and, in particular: (1) scarcity of able, enthusiastic researchers; (2) lack of economic resources to be spent and poor quality of judgment when it is spent; (3) absence of solid social development policy. Focus on Mexico and experiences of Spain.

862. Murguerza, Daniel, Daniel Bouille, and Erik Barney. 1990. "A Method for the Appraisal of Alternative Electricity Supply Options Applied to the Rural Area of Misiones Province, Argentina." ***World Development*** **18: 591-604.**

Looks at an Argentine research effort that was intended to develop a method for determining how to meet rural electricity needs as inexpensively as possible. This involves an attempt to develop non-conventional and decentralized systems on equal terms with more conventional grid-based methods of planning rural energy systems. These combined methods are applied to the province of Misiones in the northeast of the Argentine Republic.

863. Nelson, J.G. and K. Drew Knight (eds.). 1985. ***Research, Resources, and the Environment in Third World Development.*** **Ontario: University of Waterloo Geography Department.**

Essays outlining the state of environmental and earth sciences in LDCs. Sections on hydrological processes, coastal development, the environmental impact of energy use on forests, cost-effectiveness of environmental management in the context of development, and the role of research in Third World countries.

864. Office of Technology Assessment. 1991. ***Energy in Developing Countries.*** **Washington, DC: U.S. Government Printing Office.**

Assesses the extent to which technology can provide the energy services developing countries need in a cost-effective and socially visible manner, while minimizing adverse environmental effects. Emphasizes how energy is supplied and used in countries and how it is linked with development and the environment. The four main topics are: energy and economic development, energy service, energy supplies, and energy use and the environment.

865. Perez-Lopez, Jorge F. 1987. "Nuclear Power in Cuba after Chernobyl." ***Journal of Interamerican Studies and World Affairs*** **29: 79-118.**

Examines Cuba's nuclear power program and the likely effects on it of Chernobyl. Finds that Cuba is heavily committed to nuclear power and that the Soviet incident is likely to have little effect. Good bibliography on Cuban nuclear power, citing a wide variety of sources in Spanish as well as English.

866. Portes, Alejandro and Adreain Ross. 1976. "Modernization for Emigration: The Medical Brain Drain from Argentina." ***Journal of Interamerican Studies and World Affairs*** **18: 395-422.**

The analysis in this article is based on interviews with young Argentine physicians planning to emigrate to the United States. It attempts to complement speculation on push and pull forces affecting the brain drain with illustration of the definitions of the situation and values held by participants in the process. Attempts to show how these subjective definitions and values are linked with broader issues of underdevelopment and theories about sources of change.

867. Posmowski, Pierrette. 1980. "Technology and Environmental Needs: The Case of India." ***Impact of Science on Society*** **30: 335-46.**

Laudatory description of India's acquisition and mastery of optimal technologies.

868. Saddayao, C.M. 1992. "Energy Investments and Environmental Implications: Key Policy Issues in Developing Countries." ***Energy Policy*** **20: 223-32.**

Policy makers determining energy investments should consider both short- and long-term costs since these investments imply changes in the physical, social, and economic environment. Reviews issues relevant to developing countries and discusses conceptual and policy issues. Suggests macroeconomic and institutional issues relevant to environmental impacts that might be incorporated in energy planning.

869. Stepan, Nancy. 1978. "The Interplay Between Socio-Economic Factors and Medical Science: Yellow Fever Research, Cuba and the United States." ***Social Studies of Science*** **8: 397-423.**

Historical study that examines the reasons for the twenty-year gap between Carlos J. Finlay's 1881 theory of the mosquito transmission of yellow fever and the 1900 confirmation of this theory by the U.S. Reed Board. Since the economically and politically powerful institutions of the US were not threatened by yellow fever prior to the US Army's occupation of Havana, little attention and funding were given to the problem.

870. Tobey, J.A. 1990. "Economic Development and Environmental Management in the Third World: Trading-off Industrial Pollution

with the Pollution of Poverty." ***Habitat International*** **13: 125-35.**

Argues that the explanation for lax environmental controls in the Third World is related to level of development and that control of environmental degradation in low-income countries is more difficult than in industrialized countries.

871. Viswanathan, P.N. and V. Misra. 1989. "Occupational and Environmental Toxicological Problems of Developing Countries." ***Journal of Environmental Management*** **28: 381-86.**

Suggests that problems of toxicology in developing countries require an approach different from that of developed countries, because of additional problems such as disease, malnutrition, overpopulation, and lack of awareness. Thc authors show how disease, malnutrition, and other complications may interact with pollution and exacerbate its effects. They maintain that occupational and environmental toxicology must be developed as a multidisciplinary science.

872. Yishai, Y. 1979. "Environment and Development: The Case of Israel." ***International Journal of Environmental Studies*** **14: 205-16.**

Maintains that Israel has characteristics of both a developed and a developing society. It is "developed" according to economic measures and "developing" according to social propensities, so that it is still "developing" with regard to environmental policies. Therefore, economic growth has not induced the active environmental politics of other developed societies. On this basis, the author suggests that a third category be added to the two-world model, to cover societies in transition from developing to developed.

Chapter 12 -- Communications

873. Ad Hoc Panel on the Use of Microcomputers for Developing Countries. 1988. *Cutting Edge Technologies and Microcomputer Applications for Developing Countries.* Boulder: Westview Press.

Proceedings of a conference jointly sponsored by USAID, the U.S. National Academy of Sciences, and Portugal's Junta Nacional de Investigacao Cientifica e Tecnologica. Consists of essays by government officials, engineers, academics, and businessmen covering four areas: The Technologies, Resource Assessment, Resource Utilization, and Communication. Describes Portugal's experiences in developing microcomputer technologies.

874. Agarwhal, Suraj M. 1985. "Electronics in India: Past Strategies and Future Possibilities." *World Development* 13: 273-92.

Growth in India's electronics production sector has been slow. Seeks to explain why local R&D is insufficient for producing internationally competitive products. Includes statistics on imports and exports of electronic goods.

875. Aina, L.O. 1989. "Bibliographic Control of the Literature of Science and Technology." *International Library Review* 21: 223-9.

Study of bibliographic control of scientific literature in Nigeria. Reviews current bibliographic activity in that country. Concludes that scientific literature in Nigeria is poorly controlled because there is no coordination of bibliographic activities. Recommends that a Bibliographical Association of Nigeria be formed. Gives lists of bibliographies, select lists of publications, indexing and abstracting lists to periodical literature, and directories of scientific research in Nigeria.

876. Ali, S. Nazim. 1989. "Acquisition of Scientific Literature in Developing Countries. 5: Arab Gulf Countries." *Information Development* 5: 108-15.

Fifth of a series of articles on acquisition of scientific literature by libraries in developing countries. Discusses the accumulation of collections of scientific literature in the six countries in the Gulf Cooperation Council. Looks at problems such as the distance of these libraries from vendors of books.

877. Altbach, Philip G. 1985. "Centre and Periphery in Knowledge Distribution: An Asian Case Study." ***International Social Science Journal*** **37: 109-18.**

India is used as a case study of knowledge production and distribution. Concludes that Third World countries can maximize their independence and autonomy within the context of the international knowledge system by national policies to concentrate scarce resources, regional cooperation, combined action to force alterations in international policies, and awareness of the problems.

878. Amarasuriya, Nimala R. 1987. "Development through Information Networks in the Asia-Pacific Region." ***Information Development*** **3: 87-94.**

Deals with the Regional Network for the Exchange of Information and Experiences in Science and Technology in Asia and the Pacific (ASTINFO), established by UNESCO. Suggests that networks such as ASTINFO perform crucially important activities by promoting general access to and use of S&T information acquired locally and regionally.

879. Baark, Erik. 1980. "Structure of Technological Information Dissemination in China: Publication of Scientific and Technical Manuals 1970-77." ***China Quarterly*** **83: 510-34.**

Analysis of 282 technical manuals on electronics and metallurgy during the mid-70s. During this period there was debate over the relative importance of training new workers from the proletariat to break the bourgeois monopoly over S&T or simply training scientists and engineers for industry to raise productivity. Books were often compiled under the local communist party committee rather than scientists working on their own and showed a preference for working-class authors.

Discusses the "two assessments" policy of the Gang of Four. Finds a decline in number of books published by individual authors. The language of translated works reveals that Russia provides most of metallurgical knowledge. Overlapping of books shows lack of centralized control over publishing. Relationships between research and production centers for the recruitment of authors. Limited number of publishers concentrated in Beijing and Shanghai.

880. Baark, Erik. 1985. "Information Policy in India and China." ***Information Development*** **1: 19-25.**

A comparative survey of the development of national information infrastructures and the formulation of national information policy in India and China. Topics include the influence of foreign models, traditions of library and information work, indigenous development, and patterns of information use. The existing organizational structures for scientific and technological information are compared. Policy implications are considered.

881. Beaumont, Jane and David Balson. 1988. "CD-ROM Technology Use in Developing Countries: An Evaluation." ***Microcomputers for Information Management*** **5: 247-62.**

Reports on an 18-month project evaluating CD-ROM technology as a medium for information delivery in developing countries. Bibliographic databases were installed in university or research institute libraries in Malaysia, Trinidad, the Philippines, Botswana, India, Cuba, and Canada. Participants in these countries responded through evaluation forms and informal interviews.

882. Benazzouz, Abderrahmane and Albert Baez. 1978. "Knowledge Transfer in Electronics: A North African Case." ***Impact of Science on Society*** **28: 329-34.**

Describes a project to equip Algeria with an institution to provide training for engineers and technicians based on the experience of a highly developed country. As a long-term project it will establish a regular exchange of teaching staff with American universities, through participation in national and international conferences, distribution of publications, offers of visiting professorships, and by maintaining close contacts between its own library and the library of the establishing consortium.

883. Beniger, James R. 1988. "Information Society and Global Science." ***Annals of the American Academy of Political and Social Science*** **495: 14-28.**

The Control Revolution is the reason much of the labor force in advanced industrial countries works at informational tasks, while wealth comes increasingly from informational goods and services. Gives an historical summary of the development of information science from mid-19th century to present. Sees a crisis of control in global systematization of science. This crisis arises from the usually implicit assumption that science primarily consists of a one-way information flow.

Information-system computerization ignores the very feedback signals by which scientific outputs are regulated, especially problematic since global science relies upon shifting networks of individual researchers. These networks must span the borders of more than a hundred countries and a wide variety of organizational and professional boundaries. Local reputations do not translate well across organizational boundaries, but status in the global system does translate well in most local contexts.

884. Boafo, S.T. Kwame. 1991. "Communication Technology and Dependent Development in Sub-Saharan Africa." Pp. 103-124 in *Transnational Communications: Wiring the Third World*, edited by Gerald Sussman and John A. Lent. Newbury Park, CA: Sage.

Examines the communication infrastructure in black African countries. Principal features are an urban orientation, poor adaptation to a rural and oral culture, and dependency on foreign sources and TNCs for almost all hardware and most software. Four reasons for this: (1) communication plans that are poorly integrated into national development strategies, (2) inadequate finances, industrial infrastructure, and skills, (3) a lack of long-range planning to produce indigenous communication technology, and (4) insufficient political will and commitment to development of a communication infrastructure.

885. Brittain, J. Michael. 1985. "The Relevance of Social Science Output Worldwide." *International Social Science Journal* 37: 259-75.

Structural barriers prevent Third World researchers from getting access to data. Argues that social scientific knowledge is not at all like natural scientific knowledge and that information services in the social sciences must take account of the special features of social scientific communication.

886. Buchner, Bradley Jay. 1988. "Social Control and Diffusion of Modern Telecommunications Technologies: A Cross-National Study." ***American Sociological Review*** **53: 446-453.**

Development and diffusion of communication technologies have proceeded in different directions in different societies. Television is an attractive tool in exposing a target population to a product, political viewpoint, or ideology. Telephone has little educational capability and is more difficult to control. Examines ideological reasons TV has spread more in Eastern European countries than telephone. U.S. is the only country in which TV is almost wholly in private hands.

887. Crawford, William B. 1990. "Information Markets, Telecommunications and China's Future." Pp. 159-170 in ***International Science and Technology: Philosophy, Theory and Policy*****, edited by Mekki Mtewa. New York: St. Martin's Press.**

Looks at the role of information markets and telecommunications in China's development. Compares three models of development--the mechanical, the organic, and the cybernetic. Mechanical models see development in terms of universal patterns of linear causation. Rostow's stages are cited as an example. Organic models also involve universal patterns of linear causation, but stress socio-psychological factors.

Cybernetic models allow for more than one set of causal factors moving toward more than one possible end. This last is most appropriate for guiding China's efforts; China must develop by participating in global information markets while minimizing information dependency. The critical role in the process of information development will be played by its "brain-intensive" talent.

888. Fadul, Anamaria and Joseph Straubhaar. 1991. "Communications, Culture, and Informatics in Brazil: The Current Challenges." Pp. 214-233 in ***Transnational Communications: Wiring the Third World*****, edited by Gerald Sussman and John A. Lent. Newbury Park, CA: Sage.**

Brazil as a case study of state intervention to preserve nationalist development with regard to communication and information technology. Examines the history of information and communication policy in Brazil and maintains that present-day problems are a result of the development model established by the military coup of 1964. Considers the origins and evolution of Brazil's National Informatics Policy (PNI) and the role of this policy in the attempt to realize technological independence. Offers suggestions for further development of communication and information policies, mostly entailing greater decentralization and democratization.

889. Galal, Salah. 1976. "Current Trends of Scientific Activity in Arab and Islamic Countries." *Impact of Science on Society* 26: 169-76.

Describes the relative success of modern methods of disseminating technical and scientific information in the Arab Republic of Egypt. A variety of different fields are covered.

890. Habermann, Peter. 1990. "Global Telecommunication Strategies for Developing Countries." Pp. 143-150 in *International Science and Technology: Philosophy, Theory and Policy*, edited by Mekki Mtewa. New York: St. Martin's Press.

Maintains that the flexibility of global telecommunication and its insensitivity to its own content enable it to replace physical infrastructures, especially in developing countries. Examines the development of the Caribbean telecommunication system and of the Caribbean Association of National Telecommunication Organizations (CANTO). Suggests that in the Third World the growth of point-to-point communication links should follow developmental priorities, but that this communication is more sensitive to foreign domination than is mass media. Continuing domination by Cable & Wireless and ITT shows that global telecommunication has not yet become a subject of critical discussion.

891. Haule, John James. 1990. "A Model for Telecommunication Development in Africa." Pp. 151-158 in *International Science and Technology: Philosophy, Theory and Policy*, edited by Mekki Mtewa. New York: St. Martin's Press.

Maintains that the division into rich and poor nations is being exacerbated by the development of information and technology haves and have-nots. Africa is hampered in its development of telecommunications by lack of trained manpower and lack of finances. Suggests that a model for telecommunication development in Africa should be two-pronged, to try to improve management and financial support for telecommunications at the same time that African countries try to adapt technological innovations.

892. Heitzman, J. 1990. "Information Systems and Development in the Third World." ***Information Processing and Management*** **26: 489-502.**

Deals with problems faced by developing nations as a result of the shift in the global economy toward development of services, including information systems. Focusing on the South Asian nations, the author suggests that the general direction of solutions in the late 1980s was away from large-scale, centralized intervention and toward more decentralized national and regional projects.

893. Ho, T.I.M and K. Sung. 1990. "Role of Infrastructure Networks in Supporting Social Values to Sustain Economic Success in Newly-Industrializing Nations." ***International Journal of Psychology*** **25: 887-901.**

Information technology can promote competitive advantage by efficiently solving internal communication problems. For example, electronic mail and computer conferences can promote consensus and democratic decision making. Recent success and smaller scale of some developing nations makes them prime candidates for infrastructure (telecommunications) networks which promote local autonomy and consensual approaches to management.

894. Hoffman, K. and H. Rush. 1980. "Microelectronics, Industry and the Third World." ***Futures*** **12: 289-302.**

Authors maintain that microelectronics are threatening the traditional export successes of the Third World, since the industries in which Third World countries have been successful, such as textiles and electronics, depend primarily on the comparative advantage of low-wage, high-skill labor. Suggests that governments must therefore intervene to obtain software capacities so that these countries will be able to utilize the new technology.

895. Kaul, Mohan. 1987. "Impact of Information Technology in

Government Systems: A Regional Overview of Asian Experience." ***Information Technology for Development*** **2: 97-131.**

Argues that the use of information technology in the public sector drives computerization in Asian countries. Outlines the concerns of research projects on the impact of technology in government, including infrastructure, organization processes, applications, human resource development, and building public awareness.

896. Kumar, R.P. and P. Attri. 1987. "Development of New Technologies in India and Their Impact on the Dissemination of Information." ***International Library Review*** **19: 387-400.**

Argues India has made great progress in creating an information technology infrastructure despite socio-economic and psychological barriers. Outlines the specific additions in such fields as computer technology, electronics, satellite transmission, telecommunications, fiberoptics, and electrophotography. Ends with an optimistic assessment of future prospects.

897. Langer, Erick D. 1989. "Generations of Scientists and Engineers: Origins of the Computer Industry in Brazil." ***Latin American Research Review*** **24(2): 95-112.**

Argues that studies searching for the reasons why Brazil succeeded in the 1970s international computer industry overemphasizes socio-political developments in that decade. The origins of Brazil's computer technology can be found 20 years earlier, when scientists eagerly pursued the "purer" sciences of nuclear and solid-state physics--which led to a related concern with microelectronics. Many failed projects in these areas set the stage for computer development by building an infrastructure of labs. Suggests this is not a path poorer LDCs can follow.

898. Maier, J.H. 1980. "Information Technology in China." ***Asian Survey*** **20: 860-75.**

Author is with the U.S. Agency for International Development. Reviews the state of Chinese information technology in 1980, including the national plan, domestic hardware and software, telecommunications, and satellite capabilities. Considers the social impact of American sales and influence through returning Chinese students.

899. Manzoor, Suhail. 1985. "Saudi Arabian National Center for Science and Technology Database." ***International Library Review*** **17: 77-90.**

Description of Saudi Arabia's S&T database. Tables show numbers of publications in Saudi Arabia in various scientific areas and the Saudi contribution to these publications. Looks at trends over time from 1960 to 1982. Saudi contributions to scientific literature published in Saudi Arabia have been steadily rising.

900. Michel, J. 1982. "Linguistic and Political Barriers in the International Transfer of Information in Science and Technology." ***Journal of Information Science*** **5: 131-6.**

Analyzes the growth in the use of English as the language of scientific literature a suggests that this is creating an "English speaking community ghetto." Proposed solutions include multilingual journals, getting English speakers to learn more languages, and development of translation activities. Tables give a good summary of linguistic distribution of S&T literature but few references.

901. Mody, Bella and Jorge Borrego. 1991. "Mexico's Morelos Satellite: Reaching for Autonomy?" Pp. 150-164 in ***Transnational Communications: Wiring the Third World,*** **edited by Gerald Sussman and John A. Lent. Newbury Park, CA: Sage.**

Looks at the various foreign and domestic relationships that influenced Mexico's decision to purchase a domestic satellite in 1985. This case study claims to demonstrate how domestic forces, notably economic (the Mexican company Televisa) and political (the Mexican state), perpetuated and reinforced dependent technological relationships with transnational manufacturers. The problem and solution of dependency do not lie in technology but in the nation's economic political alliance.

902. Moll, Peter. 1987. "Information Technology in the Caribbean." ***Information Development*** **3: 95-102.**

Describes the work of agencies such as UNESCO in the development of regional information networks in the Caribbean and suggests that a systematic review is needed of opportunities and methods for librarians in the region.

903. Nilsen, Svein Erik. 1979. "The Use of Computer Technology in Some Developing Countries." ***International Social Science Journal*** **31: 513-28.**

Analyzes the history of computers and their use in some developing countries in order to balance the information diffused by the computer companies. Suggests that if used unwisely computers can have a negative impact on employment and on society as a whole.

904. Oeffinger, John C. 1987. "Merging Computers and Communication: A Case Study in Latin America." ***Telematics and Informatics*** **4: 195-210.**

Case study of a microcomputer-based international network that involves institutions in Brazil, Chile, Costa Rica, Mexico, Uruguay, Venezuela, and the U.S. Intended to answer questions regarding the linking of computers and communications on individuals in developing countries. Describes training of participants, problems, initialization of communications/pilot projects, development of e-mail capabilities and model educational conferences, development of scientific search capabilities, additional training, cost considerations, and future expectations.

Individuals and organizations that gain access to international electronic networks cannot expect immediate results; attention must focus on applying technology to existing networks in order to achieve an early positive impact. Cost is a major barrier, but organizations can share computer units to reduce cost. Barriers to individual information access can be reduced by clearinghouse organizations that make information readily available. Because low capital is needed to access electronic information, as modems surface in developing countries, opportunities for information-age participation will increase.

905. Ogan, Christine. 1988. "Media Imperialism and the Videocassette Recorder: The Case of Turkey." ***Journal of Communication*** **38: 93-106.**

Suggests that the media imperialism thesis which arose in the mid- to late-70s should be rethought as a result of the increasing decentralization of media, particularly through the use of VCRs. Looks at the economic and cultural components of this thesis and suggests that both dimensions presume a centralized approach to development and distribution of media products. A case study of Turkey shows that technology may be used in more and more culturally specific ways. Some good references on media and cultural domination.

906. Paez-Urdaneta, Iraset. 1989. "Information in the Third World." *International Library Review* 21: 177-91.

Deals with the problems of information for development in the Third World. Suggests that the notion of the New International Economic Order (NIEO) has failed and that the two main international initiatives on information (UNISIST and NATIS) have lost their power. Suggests information strategies for Third World countries, including the development of a National Information Policy. Few references.

907. Pagell, R.A. 1990. "New Information Technologies in Libraries and Information Centers in Third-World Countries." *Online* 14: 100-101.

Report on responses to a questionnaire given by the author to 36 members of the 1989 three-month training course at the Asian Institute of Technology in Bangkok. Respondents from 14 different countries in Asia. Author concludes by raising questions about the use of machine-readable data outside the West.

908. Schiller, Herbert. 1976. *Communication and Cultural Domination*. White Plains, NY: International Arts and Sciences Press.

Argues direct colonization of Third World nations has been replaced by cultural domination. Media technologies left behind by former colonial rulers, as well as technologies imported by fledgling nations, are not culturally neutral. They shape communications within LDCs in the mold of core nation needs rather than developing nation needs. Includes a critique of the "right to free speech" as a bourgeois ideology in the best interest of the capitalist countries but not necessarily beneficial for the periphery.

909. Seusing, Ekkehart. 1989. "The Importance of Publications from Developing Countries and the Implications for Libraries in Industrialized Countries." *IFLA Journal* 15: 118-27.

Account of the importance of publications from developing countries, with emphasis on the soft sciences. Among the conclusions: (1) demands for publications and especially statistics produced in developing countries are determined by exogenous factors, (2) special and research libraries act only as intermediaries between producers and users.

910. Sussman, Gerald. 1987. "Banking on Telecommunications: the World Bank in Philippines." ***Journal of Communication*** **37: 90-105.**

Analyzes World Bank activity and goals in developing a telecommunications infrastructure in the Philippines. As part of the international economic system, it benefits foreign corporations more than the local population. References consist largely of World Bank documents and of leftist critics of MNCs in the Philippines, such as Renato Constantino and Walden Bello.

911. Sussman, Gerald and John A. Lent (eds.). 1991. ***Transnational Communications: Wiring the Third World.*** **Newbury Park, CA: Sage Publications.**

Thirteen papers on international communications that take the perspective of the New World Information and Communication Order (NWICO) and oppose what the authors see as media and data monopolies of the West. These articles discuss the political economy of Third World communications and the attempts of peripheral and semi-peripheral countries to escape their marginalization.

912. Sussman, Gerald. 1991. "Telecommunications for Transnational Integration: The World Bank in the Philippines." Pp. 42-65 in ***Transnational Communications: Wiring the Third World,*** **edited by Gerald Sussman and John A. Lent. Newbury Park, CA: Sage.**

Uses a case study of the Philippines to look at the role of the World Bank in promoting telecommunication transfers to the Third World. Using a critical, political economy perspective, the author argues that the World Bank has furthered interests of the TNCs, supported the Marcos dictatorship, and tightened its control of the Philippines under Aquino. The national communication policy of the Philippines is seen as an expression of the export-oriented industrialization pushed by the World Bank.

913. Sussman, Gerald. 1991. "The 'Tiger' from Lion City: Singapore's Niche in the New International Division of Communication and Information." Pp. 279-308 in ***Transnational***

***Communications: Wiring the Third World*, edited by Gerald Sussman and John A. Lent. Newbury Park, CA: Sage.**

Considers the NICs as semiperiphery in a new international division of labor, with specialized functions in the world economy. Looks at Singapore in this context, seeing its information and communication industries as government dominated and closely tied to the interests of TNCs. Singapore's economic success is due to its special niche and therefore not applicable to other developing countries. The communications infrastructure in Singapore is seen as supporting the repressive government of Lee Kwan Yew.

914. Wad, Atul. 1982. "Microelectronics: Implications and Strategies for the Third World." *Third World Quarterly* 4: 677-97.

Considers how microelectronics technology should be interpreted in light of the historical experiences of the Third World with technologies from advanced countries and the current political economic climate of development. Includes discussion of the ways the effects of this industry in the Third World have been examined by other authors and suggests that most have been enthusiastic about potential benefits for the South. The author is critical of the view that microelectronics offer a panacea for problems in developing countries and feels that it could widen the gap between North and South. South-South cooperation is chief among the strategies suggested.

915. Westman, John. 1985. "Modern Dependency: A 'Crucial Case' Study of Brazilian Government Policy in the Minicomputer Industry." *Studies In Comparative International Development* 20(2): 25-47.

Shows how Brazil was able to keep control over its minicomputer industry despite heavy intervention by U.S. corporations, thus challenging dependency theory.

Chapter 13 -- Science

916. Basalla, George. 1967. "The Spread of Western Science." *Science* 156: 611-622

Historical model for the introduction of modern science into non-European nations. Three overlapping phases: (1) nonscientific society provides a source for European sciences, (2) a period of colonial science, and (3) the process of transplantation completed with a struggle to achieve an independent scientific tradition.

917. Buck, Peter. 1975. "Order and Control: The Scientific Method in China and the United States." *Social Studies of Science* 5: 237-67.

Historical study of the Science Society of China, formed in Ithaca, New York. Brief history of scientific ideas in the late Qing dynasty. American-trained Chinese scientists linked progress of science with national transformations but abstracted American views of science and its methods from industrializing concerns of 19th century America. Chinese looked to science for integrative ideas in the midst of the breakdown of Qing social order.

918. Dedijer, S. 1963. "Underdeveloped Science in Underdeveloped Countries." *Minerva* 2: 61-81.

One of the first articles to emphasize institutional barriers to "scientific development" in the Third World--the lack of a scientific community with its own traditions, the neglect of the importance of research by the Third World industrial sector, and the centralization of economic decision making by a government lacking experience in dealing with science policy.

919. Eisemon, Thomas O. 1979. "The Implantation of Science in Nigeria and Kenya." *Minerva* 17: 504-26.

Outlines the relative success of scientific communities in these two African nations, despite ethnic conflicts and an emphasis by researchers on career advancement and status rather than professional achievement.

920. Fuenzalida, Edmundo F. 1983. "The Reception of 'Scientific Sociology' in Chile." ***Latin American Research Review*** **18: 95-112.**

Historical examination of the development of scientific sociology in Chile. This approach has fallen into disfavor because of its inappropriateness to Chilean social realities and because it has served as a link to international centers of capitalism, rather than as a tool of liberation.

921. Gomezgil, Maria Luisa Rodriguez Sala de. 1975. "Mexican Adolescents' Image of the Scientist." ***Social Studies of Science*** **5: 355-61.**

Results of a national study of students in state and private secondary and preparatory schools. Describes positive and negative composite images of scientists, finding that schools have inculcated universal images of the scientist. In dependent countries like Mexico these images should be modified by "local characteristics."

922. Herzog, A.J. 1983. "Career Patterns of Scientists in Peripheral Communities." ***Research Policy*** **12: 341-9.**

Tests for efficacy of a number of possible linking mechanisms between peripheral community scientists and colleagues abroad. Shared work experiences in a foreign organization are most likely to be productive in facilitating later communication between peripheral and core scientists.

923. Hess, David. 1987. "Religion, Heterodox Science and Brazilian Culture." ***Social Studies of Science*** **17: 465-77.**

Discusses parapsychology in Brazil and its adaptation to the social and cultural institutions already present in that country.

924. Mendelssohn, Kurt. 1976. ***Science and Western Domination.*** **London: Thames and Hudson.**

The method of science is its essence: the Cultural Revolution in China will inculcate this method in the thought of ordinary people. Argues that the importance of science to development is limited to the physical sciences.

925. Mohseni, Manouchechr. 1976. "Sociological Research in Iran." ***International Social Science Journal*** **28: 387-90.**

Historical overview of the development of Iranian sociology and a description of the present state of the discipline (written before the Iranian revolution).

926. Monkiewicz, Jan. 1989. "Determinants of National Technological Performance." *Science of Science* 6: 247-60.

Outlines major theoretical perspectives, points out their weaknesses, and proposes a synthesis.

927. Moravcsik, M.J. 1986. "Two Perceptions of Science Development." *Research Policy* 15: 1-11.

Two contrasting views of the role of science development. One emphasizes the value of science in terms of short-term economic growth; the other considers the value of science in a broad and farsighted context. Adherence to the first view is responsible for the slow development of science in the Third World.

928. Moravcsik, Michael. 1989. "Dependence and Science Scenarios for the Third World." *Social Science Information* 28: 445-52.

Six meanings of "dependence" with respect to S&T, followed by 5 "scenarios" for the Third World. All countries are "dependent" on others for world science.

929. Needham, Joseph. 1981. *Science In Traditional China: A Comparative Perspective.* Cambridge: Harvard University Press.

Technology in China was more advanced than in the West until 16th century. Lectures on gunpowder and firearms as they developed from alchemy, comparative macrobiotics, acupuncture, and attitudes toward time and change together with some anecdotes about the massive series Science and Civilization in China.

930. Needham, Joseph and C.A. Ronan. 1990. "China and Europe: Their Different Progress in Science." *Interdisciplinary Science Reviews* 15: 301-9.

By the foremost Western scholar of the history of Chinese science. In this article, the authors discuss why modern science arose in the West and not in China. Among these reasons are the attraction of the best Chinese intellects to the bureaucracy and a continuation of the traditional order of things which did not encourage radically new thinking. Despite this, the West owes many technological advances to China.

931. Omahony, P. 1991. "Science and Industry." ***Education and Training Technology International*** **28: 30-42.**

Maintains that industry loses its autonomy in modern society as science dictates the economic agenda. Criticizes both the Marxist and technocratic explanations of the relationship. Suggests that a sociological learning theory articulated in the form of a theory of collective learning is necessary for an explanatory account of the present science-industry relationship.

932. Orleans, Leo (ed.). 1980. ***Science in Contemporary China.*** **Stanford: Stanford University Press.**

Chapters on science in China's past, science policy and organization, then special chapters on each discipline.

933. Pattnaik, Binay Kumar. 1989. "Scientific Temper and Religious Beliefs." ***Journal of Sociological Studies*** **8: 13-40.**

Finds little correlation between religious belief and ability to perform science, suggesting that the two may be partly hostile but not antithetical in India.

934. Rabkin, Yakov M. 1986. "Cultural Variations in Scientific Development." ***Social Science Information*** **25: 967-989.**

Looks at ways to understand comparative development of scientific cultures in different peripheral countries. Suggests that these countries may be divided into two groups: those whose cultures have historically descended from or acquired a commitment to Western European roots (such as most Latin American countries, Poland, and Greece) and those of non-Western origin who integrate Western elements as an instrumental concession or a necessary evil. The paper argues that cultural allegiance may be a prime factor of success in the development of science. The author finds that attempts to create alternative sciences, "liberated" from domination of Western values, have often degenerated into an obsession with being the reverse of the hypothetical model of Western science.

935. Salam, A. 1966. "The Isolation of the Scientist in Developing

Countries." *Minerva* 4: 461-5.

A compassionate account of reasons why scientific research lags in developing countries--social pressures for those with intellect to enter the bureaucracies, the small number of appropriate role models to educate, cultural emphasis on moral rather than academic achievement.

936. Salomon, Jean-Jacques and Andre Lebeau. 1990. "Science, Technology, and Development." *Social Science Information* 29: 841-858.

Final article in a symposium on science and technology development, which originated in response to Salomon and Lebeau's book *L'Ecrivain public et l'ordinateur--Mirages du developpement*. The authors comment on the contributions to the symposium. They defend their position that basic research has no direct impact on the economic development of the Third World. History and culture are important for an understanding of technological take-off in the NICs, or technological lag in other countries. They clarify and repeat their position that developing countries must selectively adopt models of industrialized countries that are appropriate to them.

937. Sardar, Ziauddin. 1989. *Explorations in Islamic Science*. London: Mansell.

Confronts the perception that Islam and science are mutually exclusive. Documents "the return of Islamic science"--science research shaped by Islamic needs, concerns, culture, and ethics. Western science is destructive and cannot satisfy needs of Muslim cultures.

938. Schneider, Lawrence A. 1982. "The Rockefeller Foundation, the China Foundation, and the Development of Modern Science in China." *Social Science and Medicine* 16: 1217-21.

Historical examination of the role of these two foundations.

939. Schoijet, M. 1979. "The Condition of Mexican Science." *Minerva* 17: 381-412.

A pessimistic account of Mexico's scientific resources, stressing that many recorded "scientists" are often part-time teachers or researchers. Relatively few Mexican scientists have doctorates or publish scientific papers. Criticizes the current allocation of research budgets and sophisticated research equipment as detrimental to innovation. Finally, points to the country's deficiency in trained technical staff.

940. Schott, Thomas. 1992. "The Scientific World System: Conceptualization." *Science, Technology, and Human Values 17.*

Network conception of the scientific world system views it as a global system of production and diffusion of scientific knowledge at the macro level, national invisible colleges at the intermediate level, and collegial environments around scientists at the micro level. Identifies important variables at each of the three levels.

941. Schwartzman, Simon. 1978. "Struggling to Be Born: The Scientific Community in Brazil." *Minerva* 16: 545-80.

Traces Brazil's relative success in scientific development to a historically favorable cultural attitude toward positivism. Brazil's periods of dictatorship have produced some scientific gains, but S&T policy has fallen under criticism as a funnel for perquisites to an elite clique. Also, Brazilian science labors under a bias toward applied research.

942. Sharma, Dhirendra. 1991. "India's Lopsided Science." *Bulletin of the Atomic Scientists* May: 32-36.

Maintains that Indian science is skewed toward nuclear, military, and space research, and has failed to deliver advances for the Indian people. The blame for this is placed on Nehru and Nehru's tsar of Indian science policy during the 1950s and 1960s, Homi J. Babha. Babha was able to direct the course of Indian science policy, and all big science research was concentrated under his own domain, the Department of Atomic Energy. Emphasis was on capital-intensive, energy-consuming nuclear technology that increased foreign dependency. Although no single man controls policy today, the narrowness and lack of originality in Indian science is the legacy of Babha.

943. Shenhav, Yehouda and David Kamens. 1991. "The 'Costs' of Institutional Isomorphism: Science in Non-Western Countries." *Social Studies of Science* 21: 527-45.

Three mechanisms that cause science in non-Western countries tends to follow Western forms. Sample of 73 LDCs shows that there is no relationship between degree of institutionalization of science and economic performance. For the least-developed countries the relationship is even negative.

944. Shiva, V. and J. Bandyopadhyay. 1980. "The Large and Fragile Community of Scientists in India." ***Minerva*** **18: 575-94.**

Dampens optimism about India's scientific community by pointing to the lack of communication among various research institutions, the lack of inventiveness shown by these institutions, the lack of a clear national research program, and constraints imposed by poor equipment availability. Indian researchers are also hindered by relatively little contact with Western scientists and the gap between bureaucratic views of research and the way science is actually done.

945. Sivin, Nathan (ed.). 1977. ***Science and Technology in East Asia.*** **New York: Science History Publications.**

This volume includes selections from ISIS on the history of S&T: including quantitative and qualitative science, technology, and cultural interaction. Essays from 1914 to 1976 on Chinese astrology, alchemy, firearms, salt mining, anatomy.

946. Suttmeier, Richard P. 1985. "Corruption in Science: The Chinese Case." ***Science, Technology and Human Values*** **10 (1): 49-61.**

Norms in Chinese science are a blend of Confucian and traditional (emulation of past scholarly achievements), political party prescriptions (since early 1950s, the use of science for socialism--quite close to Mertonian norms), and those of the work unit (danwei, which controls the career and is often at odds with party policy). The goals and norms of the Party are consistent with Mertonian scientific norms in ways that the bureaucracy of the danwei is not. Example of "beancurd" processing as a case of bureaucratic insubordination. Hypothesizes that the more important S&T is to upward social mobility in a society, the greater the motivation to deviate (more so in China than in U.S., where many other avenues exist).

947. Thein, Mya Mya. 1979. "Women Scientists and Engineers in Burma." ***Impact of Science on Society*** **29: 15-22.**

Suggests women have full equality in the sciences in Burma as a result of the newness of its research community.

948. UNESCO. 1985. *Science and Technology in the Countries of Asia and the Pacific*. Paris: UNESCO.

Discusses general features of Asian and Pacific countries, their science and technology framework and potential for S&T development. Bibliographies for each country or region.

949. Watson, Helen. 1990. "Investigating the Social Foundations of Mathematics: Natural Number in Culturally Diverse Forms of Life." *Social Studies of Science* 20: 283-312.

Comparison of English and Yoruba methods of predication showing their impact on the kinds of objects they have and alternative constructions of number; Yoruba numbers are verbal constructs, rather than adjectival.

950. Wen-Yuran, Qian. 1982. "The Great Inertia: An Introduction to a Causal Inquiry into Traditional China's Scientific Stagnation." *Comparative Civilizations Review* 9: 23-44.

Author from Zhejiang University disagrees with Joseph Needham's admiration for traditional Chinese science. Sees the ideological and political unification of dynastic China as the source of China's inertia.

951. Ziadat, Adel A. 1983. *Western Science in the Arab World: The Impact of Darwinism, 1860-1930*. London: Macmillan Press Ltd.

Case study of the Muslim and Christian response to Darwinism, both academic and cultural, in the Arab world. Includes discussions of the debate over materialism and general Arabic reaction to Western culture.

Chapter 14 -- Development of Science and Technology

952. Abdallah, Abdel-Aziz Ibn. 1976. "Problems of Arabization in Science." *Impact of Science on Society* 26: 151-60.

Suggests the inflexibility of modern Arabic is an obstacle to Arab scientific development.

953. Abiodun, A. Ade. 1981. "Technology Development in Africa." *Africa* 113: 56-7.

An essay that argues against the tendency for African nations to "overgorge" on technology. Suggests that Africa halt its global search for "a panacea to our unique problems and look to the domestic scene for innovations." Training programs should be geared to African needs rather than foreign standards.

954. Ahmad, Aqueil. 1986. "Western Science and Technology in Non-Western Cultures." *Science and Public Policy* 13: 101-5.

Policy conflict exists in India and other Asian nations over social and scientific progress. Western science in Indian cultures produces contradictions. Education and social policy aimed at eliminating cultural deprivation are needed. S&T advances are hampered by a preindustrial mentality and the Indian ability to exist happily in spite of poverty.

955. Bagchi, Amiya Kumar. 1988. "Technological Self-Reliance, Dependence and Underdevelopment." Pp. 69-91 in *Science, Technology and Development*, edited by Atul Wad. Boulder: Westview Press.

Theoretical overview of the concept of self-reliance, ways of achieving it, impediments to it, and stages in the attainment of self-reliance. The historical roots of dependence are traced to colonialism. The ability of a country to attain self-reliance is said to depend on (1) the size of the country, (2) the country's geopolitical situation, (3) the character and extent of changes in technology and (4) the state and direction of capital flows in the world economy.

Three major routes to technology absorption and adaptation are discussed: (1) the profit motive, (2) the "associationist route" (basically cooperative arrangements between enterprises), and (3) political mobilization (such as in a socialist economy or wartime capitalist economy). Examines technology diffusion in the attainment of self-reliance, and suggests that diffusion can have a beneficial effect on both supply and demand if it raises both productivity and incomes.

For grasping the economic characteristics of a global economy, the author offers a four-fold classification, cross-tabulating high or low economies of scale with high or low rates of technical progress. The greater the degree of economies of scale in a frontier technology, the harder it is for a Third World country to master it. Sees the basic impediments to the attainment of self-reliance as political and social.

956. Beranek, William and G. Ranis (eds.). 1978. *Science, Technology, and Economic Development: A Historical and Comparative Study*. New York: Praeger.

A collection of articles surveying the historical role of science and technology in the development of Brazil, Ghana, Hungary, Britain, modern Germany, and the United States. The final article points to the "gaps" between S&T and development, arguing that social institutions must adjust to facilitate the diffusion of major technological innovations, as must patterns of family life and work, placing particular emphasis on the importance of human resources.

957. Bhagavan, M.R. 1990. *The Technological Transformation of the Third World*. London: Zed Books.

Looks at the worldwide impact of technology over the past 150 years. Argues that modern technology was the third force, in addition to politics and economics, that made it possible for Western capital to enter every part of the world. The first chapter gives a good overview of the present technological situation in the Third World, using nine indicators of industrialization. Chapter 2 gives strategies for technological advance. The conclusion urges Third World countries to avoid the environmentally destructive forms of development of the developed capitalist and socialist worlds. The Annex Tables give excellent statistical summaries, with information drawn from *UN Statistical Yearbooks*, World Bank reports, and other sources.

958. Childers, E. 1979. "Technical Co-operation Among Developing Countries: History and Prospects." ***Journal of International Affairs*** **33: 19-42.**

Cooperation among Third World countries was an idea developed at a UN-sponsored conference in 1978 and described in this participant's account. Gives a brief history of technical cooperation from ancient times through colonial, in which attitude of total technical dependency upon the North prevailed. The view that LDCs have something to offer each other evolved gradually.

959. Cohen, R.S. 1982. "Science and Technology in Global Perspective." ***International Social Science Journal*** **34: 61-70.**

Fusion of craft technology and science is recent. It is unique to western European civilization. Discusses discontinuities and failures of science, recent technological developments of world significance, elitism, and technological threats.

960. DeGregori, Thomas R. 1985. ***A Theory of Technology: Continuity and Change in Human Development.*** **Ames: Iowa State University Press.**

Challenges the views that technological progress is detrimental. Technology creates resources and fewer problems than generally thought. Considers the role of technology in human evolution and proposes 30 basic "principles" of technology to provide grounds for technology choice. Critical of Appropriate Technology and compares the dissent against technological progress to apocalyptic theories. Empirical focus of the book is on issues of agriculture, population, and nutrition.

961. Faruqui, Akhtar Mahmud. 1986. "Science and Technology: The Third World's Dilemma." ***Impact of Science on Society*** **36: 3-14.**

Historical discussion by a Pakistani author, examining the prospects for change in the "developed" world's dominance over big science and high technology. Why science and technology in LDCs is so behind and reasons why this trend might improve soon.

962. Galtung, Johan. 1979. "Towards a New International Technological Order." Pp. 277-300 in ***Science, Technology and the***

***Social Order*, edited by Ward Morehouse. New Brunswick, NJ: Transaction Books.**

Holds that technology is not merely a neutral mode of production, but that it carries economic, social, and cultural codes. The economic code of Western technology requires that industries be capital-, research-, organization-, and labor-intensive. The social code perpetuates inequality by creating a center and a periphery. The cultural code sees the West as destined to recreate the world in its own image. The cognitive code conveys an image of man as master of nature, vertical and individualistic relations between humans as normal, and history as a linear movement of progress. These structural codes determine which techniques will be researched, developed, and deployed. For techniques that induce different structures, an awareness of the interplay between techniques and structures and political will are necessary.

963. Ghazanfar, S.M. 1990. "Third World Technological Change: Some Perspectives on Socioeconomic Implications." *Journal of Social, Political, and Economic Studies* 15: 91-105.

Social innovations such as institutional, attitudinal, and behavioral adjustments are necessary to accomplish modern technological innovations. Offers a very general discussion of the meaning of S&T and technological change, links between technological and social change, and effects of technological change on social structure.

964. Ghazanfar, S.M. 1980. "Individual Modernity in Relation to Economic-Demographic Characteristics: Some Evidence from Pakistan." *Studies in Comparative International Development* 15: 37-53.

Studies the relationship between social/cultural factors and national development, including attitudes toward science.

965. Goonatilake, Susantha. 1988. "Epistemology and Ideology in Science, Technology and Development." Pp. 93-114 in *Science, Technology, and Development*, edited by Atul Wad. Boulder: Westview Press.

Examines the social epistemology of Western science. How European ideologies have affected the development of a Eurocentric science. In the literature on the contextual nature of science and knowledge, the author sees three levels of theories of scientific change: (1) theories of long-range, macro historical change (Needham, Hill), (2) intermediate-level theories, that deal with how science is governed by national policies, economic criteria, and pressure groups, and (3) micro-level theories of how social factors structure science for particular communities of scientists.

Since knowledge is selected and legitimized depending on how well it fits with social criteria, different historical social circumstances could have provided different scientific approaches and different methodologies. This suggests that there could be an alternative to the science that developed under European hegemony. Discusses science in India as an example of a "dependent knowledge structure". Gives views on how to attain a creative science in a Third World context. Suggests that detaching a country from the international base of science (as in the Chinese Cultural Revolution or Khmer Rouge Cambodia) may not be successful. Instead, she suggests "transcending" the linkages and developing alternative scientific social linkages.

966. Gvishiani, Dzhermen M. 1980. "Development Problems, Contemporary Science, and Technology." *International Social Science Journal* 32: 151-7.

Developing countries have not been very successful in improving their technological capacities. Estimates that from 35% to 45% of the world's scientists and specialists are directly or indirectly engaged in research and development connected with military matters.

967. Inkster, Ian. 1985. "Scientific Enterprise and the Colonial 'Model': Observations on Australian Experience in Historical Context." *Social Studies of Science* 15: 677-704.

Outlines spread of modern science to Australia, and suggests the process reveals a loosening of established imperial relations.

968. Kettani, M. Ali. 1976. "Moslem Contribution to the Natural Sciences." *Impact of Science on Society* 26: 135-50.

Author maintains that we do not owe inductive reasoning or modern experimental science to western civilization. They are a continuation of what Muslims and Arabs began during the Renaissance. The Koran prescribes an approach to research that is basically inductive. The article is highly tendentious, but provides useful historical background.

969. Lall, Sanjaya. 1975. "Is 'Dependence' a Useful Concept in Analysing Underdevelopment?" *World Development* 3: 799-810.

Argues that the concept of dependence is impossible to define in terms of static or dynamic criteria. Claims that dependence is usually given an arbitrarily selective definition which picks certain features of the broader phenomenon of international capitalist development. The selectivity tends to misdirect research in the area of underdevelopment. Concludes that the dependence model must be severely qualified in order to remain in use in the study of underdeveloped countries.

970. Lengyel, Peter (ed.). 1981. *International Social Science Journal* 33: 431-522.

Issue on "Technology and Cultural Values." Includes Technology Assessment, comparative interaction within Science, Technology and Society, technological self-reliance and technology choice among its topics.

971. Moore, Wilbert E. 1979. *World Modernization: The Limits of Convergence*. New York: Elsevier.

The chapter entitled "technification" puts the adoption of technology within the perspective of modernization. The development of the factory system of industrial production permits and requires technical features of "rationalized fabrication." Discusses alternative technologies but sees the primary characteristic of "technification" in highly rationalized economies as increasing capital-intensivity, and this characteristic may be undesirable in developing countries for economic, social, and political reasons. Gives a brief overview of social and environmental prices of technification.

972. Morehouse, Ward. 1979. "Science, Technology, Autonomy, and Dependence: A Framework for International Debate." Pp. 387-412 in *Science, Technology and the Social Order*, edited by Ward Morehouse. New Brunswick, NJ: Transaction Books.

Considers technology as an instrument of control on the part of the North. Sets forth issues for international debate at conferences: (1) technology as the new instrument of global domination, (2) the role of technology in the accelerating de-industrialization of the South by the North, (3) technological autonomy, not capacity, as the goal of developing countries, (4) selective technological delinking as a strategy for achieving greater autonomy, (5) technologies for basic human needs and the imperialism of social priorities, and (6) the new development myth of technological change to alleviate poverty in the absence of a social transformation. Gives some ideas for action at the national level in Third World countries, action at the national level in industrialized countries, and action at the international level.

973. Rahman, A. 1981. "Interaction Between Science, Technology and Society: Historical and Comparative Perspectives." *International Social Science Journal* 33: 508-21.

Interaction of science and technology with society may be seen as a part of development of European society and also in terms of impact of developments there on Asia. Historical discussion of technology in European and non-European cultures. In Asia (especially India), the major impact of these developments was the creation of two rival sectors: one based on handicrafts and another based on imported technology. Concludes that the interaction of technology and society depends on (1) whether the former has taken roots in society, (2) whether technology has increased or decreased inequality and injustice, (3) relationships of society with international forces.

974. Rahman, A. 1984. "Science and Technology in Indian Culture." *Bulletin of Science, Technology and Society* 4: 402-4.

Outlines a research program into the history of science and technology sponsored by the National Institute of Science, Technology and Development Studies. In particular, the program emphasizes India's contribution to world science prior to the "introduction" of science by the British. Calls for increased awareness of the interaction between S&T and cultural values, as well as between S&T and the economic system.

975. Sagasti, Francisco R. 1980. "Towards Endogenous Science and Technology for Another Development." *Technological Forecasting and Social Change* 16: 321-30.

Historical survey of development of S&T in Third World. Argues for strengthening autonomous capacity.

976. Sardar, Ziauddin. 1988. *The Revenge of Athena: Science, Exploitation and the Third World.* London: Mansell Publications Ltd.

Argues that "science perpetuates violence against the people, societies, economies, environments, traditions, cultures, ontologies, and epistemologies of the Third World." Questions "what possibilities the Third World can itself develop to meet the challenge of Western science." Twenty-one essays on topics such as health, hunger, sex and race, atomic energy, the Green Revolution, Islamic and Indian science, and AT.

977. Urevbu, A.O. 1988. "Science, Technology and African Values." *Impact of Science on Society* 151: 239-48.

How the individual African tends to view the use of science and technology in everyday life, and the cultural meaning it holds for the continent. Relationship between African cultural values and the development of S&T, accentuating the importance of actively choosing technology responsive to Africa's needs. Competing approaches: (1) no place in Africa where the potential is lacking, (2) modern science and superstition share common grounds--no basic difference in modes of thoughts, and (3) important differences between Western and African technologies.

978. Wad, Atul (ed.). 1988. *Science, Technology, and Development.* Boulder: Westview Press.

Contains 14 papers that provide perspectives on different aspects of the relationship between science and technology and the development process. General topics includes innovation, new technologies, and historical, cultural, and regional perspectives. Final section on policy and management approaches as well as indicators.

PART IV -- Indexes

Author Index

Numbers refer to entries, not pages, except where indicated. Includes essays contained within books edited by the scholar.

Subject Index

Index references indicate entry numbers, not page numbers.

Part V -- About the Authors

WESLEY SHRUM has been on the faculty of the Department of Sociology at Louisiana State University since 1982. He has served as secretary of the Society for Social Studies of Science since 1987. His book, *Organized Technology: Networks and Innovation in Technical Systems*, was published by Purdue University Press. A review essay with Yehouda Shenhav related to the topic of the present volume is published in the *Handbook of Science, Technology, and Society* (Sage, 1994). His current project is an assessment of research capacity in Ghana, Kenya, and India (Kerala), focusing on agriculture and the environment.

CARL L. BANKSTON III (BS, sociology, Southern Methodist University; MA, history, UC Berkeley) is currently a doctoral candidate at Louisiana State University. Throughout the 1980s, he worked in developing countries in Southeast Asia. He has published a number of articles on Third World minority groups in Asia and in the United States, and is working on a book about young Vietnamese immigrants, entitled *Growing Up in America*.

DENNIS GEORGE (STEPHEN) VOSS JR. is currently a doctoral candidate in the Government Department of Harvard University's Graduate School of Arts and Sciences. A specialist in political methodology, his projects have covered a range of topics including political communication, international governmental organizations, public opinion polling and racial politics in the U.S. South. Voss also edited the 1992 edition of *Let's Go: USA* (St. Martin's Press), and wrote numerous newspaper articles as a reporter with Gannett News Service.